AIR PRODUCTS :-

GW00600880

Process Fan and Compressor Selection

Process Fan and Compressor Selection

IMechE Guides for the Process Industries

Edited by

John Davidson, MBA, BSc (Hons), CEng, FIMechE

and

Otto von Bertele, BSc (Hons), CEng, MIMechE

Mechanical Engineering Publications Limited, London

First published 1996

ISBN 0 85298 825 7

A CIP catalogue record for this book is available from the British Library.

Printed in Great Britain by Antony Rowe Ltd, Chippenham, Wiltshire

Contents

Preface

Pumps, fans and compressors provide the heart and lungs of process plants. Because they are dynamic items of equipment, they also provide most of the engineering challenge, and a fair proportion of the problems associated with process plants. The reliability of such equipment is, therefore, of paramount importance to the petroleum, petro-chemical, chemical and gas industries, as well as other process and service industries. High reliabilities can only hope to be achieved if the most appropriate choice of equipment, to cover all the various process conditions, is made initially. So the process of selection of pumps, fans and compressors merits a great deal of attention in order to make the most appropriate choice. The selection of process pumps has already been dealt with in this series of IMechE Guides for the Process Industries (1) and the intention here is to deal with the Selection of Fans and Compressors.

As for process pumps, there are no British national specifications which lay down design and construction requirements for fans and compressors, and the Machines Engineer does not have the benefit of British Standards to guide his endeavours. The oil and petro-chemical industries base their procurement of fans and compressors on the American Petroleum Institute Standards API 672 (2), API 617 (3), API 618 (4), API 619 (5) and API 672 (6), and the remainder of the process industries can base their procurement on the ISO Standards ISO 13707 (7), ISO 10439 (8) and ISO 10440 (9). Notwithstanding this, and even where API Standards and ISO Standards are used, there is still the vital earlier stage in the equipment procurement procedure, i.e. the selection of the machines to meet the specified process duties. In this respect most fan and compressor manufacturers have published articles on, or have available in their sales literature, guidance on how to select equipment. Closer examination of this material reveals a number of quite understandable snags however. This manufacturer's guidance begs the first question of what options are available to the user. It assumes that the specific manufacturer's equipment has already been selected; the guidance is limited to the frame sizes available from that manufacturer; and ignores the important effect on the selection of equipment of comparative reliability and cost. Unfortunately there is little or no help on this subject in the published literature. Further, even for those industries which make use of the API Standards this does not solve all the engineer's problems as no guidance is given in the vital area of how the fans and compressors are to be integrated into the complete process system.

The strength of this publication is that it represents the distilled experience of both users and plant designers in a number of companies in the oil, petrochemical, chemical and engineering contracting industries. It is a classic example of experimental learning, where design and operating experience with hundreds of fan and compressor installations has been categorized, and analysed, and the essential lessons extracted. These essential lessons have then been synthesized to provide a series of empirical formulae, design guides and codes of practice for fan and compressor users. It is the very essence of professional engineering practice.

The guide is aimed at those engineers and managers, either of mechanical or chemical engineering disciplines, who are involved in the specification, selection and procurement of fans and compressors and yet do not claim to be experts. For other more experienced engineers and managers, it should prove an invaluable memory aid for the selection process. In addition it will provide equipment manufacturers with a great deal of insight into the problems facing fan and compressor users and plant designers. It explains how uncertainties which arise during the plant design process are dealt with when decisions need to be taken on fan and compressor selection.

The publication is also aimed at the training of young engineers who choose a career in the process industries. Whether they are finally to end up as project engineers or project managers, or take up more specialist roles, an essential part of the training of young engineers is how the selection of plant and equipment is carried out. This applications engineering, which requires a fundamental understanding of the equipment in question, and of the plant system of which it forms a part, is of prime importance in the training of engineers for the process industries.

The units used throughout this guide are SI Units.

Acknowledgements

The material comprising the guide was provided by a Working Party set up by the Process Industries Division of the Institution of Mechanical Engineers. The composition of the Working Party is as follows:

J Davidson (Chair)	Consultant	Editor
I M Arbon	Caledonian Compressors Limited	Author
O von Bertele	Consultant	Author/Editor
P J Critchley	BP Research & Engineering	Author
K G Rayner	ICI Engineering	Author

Thanks are due to the members of the Working Party for all the time and effort which went into the preparation and discussion of the base material, particularly as they are all very busy people and work on the guide has had to be fitted into already full schedules.

The Editors also wish to express their appreciation to Mr Trevor Cuerel of BP Research and Engineering (retired) and Mr Roy Jakes of Davy Process Technology, who were members of the Working Party in the early days, for all their work and effort put into the publication.

Gratitude is also expressed to the Directors of ICI Plc for permission to use in-house ICI material in the preparation of Chapter 3: Fans and Chapter 6: Reciprocating Compressors of the guide.

The Editors also wish to thank Shell Internationale Petroleum Maatschappi BV for permission to view the relevant Shell documents which greatly helped the Working Party with their deliberations.

Biographical Notes

John Davidson

John Davidson graduated as a Mechanical Engineer, and then spent three years studying in the United States, first at the University of Kansas and then at the Massachusetts Institute of Technology, where he was awarded a Masters Degree in Industrial Management. He was commissioned in REME on his return to the UK and spent two years in Germany. He then joined ICI Petrochemicals Division and undertook a variety of senior engineering roles including managing the Power Station Design Section and the Machines Design Section. He then transferred to Pharmaceuticals Division (now Zeneca) as Works Chief Engineer and then later as Group Engineer in the Division Engineering Department, responsible for all technical design activity. He left ICI and worked for a short time in a Polytechnic and then set up his own consultancy company with two colleagues. He has edited two books in the current series, one of which has been published in Italian.

Ian Arbon

After training on compressors and turbines with Peter Brotherhood Ltd, Ian Arbon worked for Delaval-Stork VOF in the Netherlands (1975) and thereafter for MAN GHH in Germany (1977), then England, and finally USA, where he became Vice-President of the Compressor and Turbine Division. He returned to Europe on joining Howden Compressors Limited (1986), where he became Managing Director. In 1991, he started his own company, Caledonian Compressors Limited (CCL), a firm specialising in the design and project management of completely packaged gas compressors (screw, vane and reciprocating) for the oil and gas process market. Ian is currently Managing Director of CCL.

Otto von Bertele

Otto von Bertele graduated as an Electrical Engineer before joining ICI Agricultural Division as an Instrument Engineer. He had various appointments on the plant and in design, including the leadership of the vibration section, before specialising on large compressors, their drivers and their lubrication systems. He has given numerous papers both at home and abroad on machines and Tribology as well as on condition monitoring and failure analysis. He left ICI when he was Company Tribologist and Machines Consultant in 1987. Since then he has been an independent international consultant with assignments in Europe, Asia and the USA.

Keith Rayner

Keith Rayner is a member of ICI Engineering which is a corporate centre for engineering good practice within the ICI Group. He has more than 20 years' experience in process plant maintenance, machine selection and specification, and machine consultancy. His specific interests are mechanical reliability, machines design and operation standards, and safety auditing.

Peter Critchley

Peter Critchley trained at Metropolitan Vickers Limited, followed by a number of years on the design of industrial, marine, and naval steam turbines. This was followed by a period with Peter Brotherhood Ltd as Chief Steam Turbine Engineer. He then moved to BP Engineering as rotating machinery engineer. He is currently a senior rotating equipment specialist with BP International Engineering Shared Service where he is a consultant on gas turbines, compressors and pumps. His work covers conceptual studies through to site troubleshooting. He is a member of the Institution of Mechanical Engineers.

Notation

Definitions

A	Area (piston, etc)	m^2
B_p	Basic pulse	N/m^2
b	Impeller width	m
C	Temperature centigrade	
c	Gas velocity	m/s
C_v	Clearance volume	percent
D	Diameter (piston, impeller etc)	m
d	Diameter (piston rod, impeller eye)	m
E	Number of stages	
H	Head	kJ/kg
L	Length of rotor, duct, etc	m
l	Length (pipe between cylinder and vessel duct, etc)	m
L_w	Sound power level	db
L_p	Sound pressure level	
Ma	Mach number	
M_w	Gas molecular weight	—
m	Mass flowrate	kg/s
N	Speed	r/s
n	Polytropic index	—
P	Power	kW
p	Pressure (absolute)	bar
Δp	Differential pressure	bar
Q	Volume flow	m^3/s
R	Gas constant 8.3143	kJ/kg mol.K
r	Stage compression ratio	
R_l	Rod load	N
S	Stroke	m
T	Absolute temperature	K
t	Temperature	C
t	Time	s
u	Tip speed	m/s

u_a	Speed of sound	m/s
U_p	Piston speed	m/s
u_v	Gas velocity in valves	m/s
V_{vol}	Volume of vessel	m^3
v	Gas velocity	m/s
w	Thickness	
w_s	Mass of suspended solids	gm/m^3
w_l	Mass of suspended liquid	gm/m^3
Y	Hub ratio (D/d)	
z	Gas compressibility	—

Subscripts

a	Actual
ad	Adiabatic
av	Average
b	Barometric
c	Corrected, compressor size
d	Delivery
dy	Dynamic
i	Inbuilt, component
in	Internal
int	Intermediate
is	Isothermal
k	Adiabatic
mech	Mechanical
max	Maximum
min	Minimum
n	Normal (1.0132 bar, 0 C)
opt	Optimum
p	Polytropic, piston
s	Suction (inlet), shaft
t	Total
v	Volumetric

Greek

α	Warm up factor	
β	Blade angle	
ε	Clearance ratio and seal coefficient	
ξ	Seal factor	
κ	Ratio of specific heats	
ρ	Gas density	kg/m^3
η	Efficiency	percent
η_v	Volumetric efficiency	percent
ψ	Pressure coefficient	
φ	Flow coefficient	
δ	Diameter coefficient	
σ	Speed coefficient	
Π	Pressure ratio	
Σ	Clearance	

Part One

Introduction and Preliminary Selection

Chapter 1

Introduction

1.1 The purpose of the guide

The primary purpose of the guide is threefold:

(a) *To identify the type of fan or compressor required to meet the specified process duty and to assemble data for the preparation of an enquiry*

The data are usually presented in the form of a Mechanical Data Sheet.

(b) *To facilitate engineering design by other members of the design team*

The iterative nature of the design process and the fact that the selection of fan and compression plant itself affects the design of the process plant, and vice versa, means that the process design must proceed hand-in-hand with the selection of the equipment. This is because the relationships of the data, both technical and commercial, are so interdependent.

(c) *To provide a basis for reviewing of offers by vendors*

It is an essential part of the equipment selection and procurement procedure that offers by alternative vendors are reviewed and comparisons of both technical and commercial features are made.

1.2 The philosophy of selection

During the early stages of design of plant for the process industries it is frequently necessary to be able to identify the type of fan or compressor required for a particular duty; to determine any constraints that it might place on the plant; and to provide estimates of certain information required elsewhere in the overall design.

Unfortunately, the design of process plant is never a straight-forward process moving forward in a simple sequential manner. On the contrary it is invariably an iterative process where assumptions are made; fed into the problem to give a partial solution; and then further improved assumptions are made which hopefully leads to an improved solution; and so on. The process of selection of fans and compressors is one stage in this process of iteration, which in itself is an iterative process.

A wide range of equipment types are available, e.g. axial and centrifugal fans; axial, centrifugal, reciprocating, screw, sliding vane and liquid ring compressors; and positive displacement blowers. For any particular process duty there may, superficially at least, be a number of possible types. The various types have different characteristics and often interact with the plant in different ways and the situation can become complex and confusing to the non-specialist.

This guide is, therefore, directed at mechanical or chemical engineers and managers, without a specialist knowledge of fans or compressors, to help them quickly, efficiently and accurately determine the information required during the preliminary design.

1.3 Design procedure

The design procedure is considered under four headings as follows:

(a) Identify the technically feasible options

The process data input is likely to be limited to suction pressure and temperature, discharge pressure, flowrate and gas composition. After processing these data would be checked against the limitations of each type of equipment to obtain a short list of types that are feasible. All types will have limitations such as maximum pressure and pressure differential, maximum discharge temperature, maximum pressure ratio, minimum and maximum flowrate and maximum power. Additionally some will be constrained by the types of gas that they can handle.

(b) Preliminary choice of equipment

This second stage would require a more formal approach to selection as shown in Part One, Chapter 2. At this stage some consideration would be given to aspects such as the reliability of the installation, maintenance characteristics, the method of control and the type of driver. In addition estimates would be made of the likely number of machines required to meet the process duty and of any installed spares, first cost and operating costs, and of size or weight constraints. Whether or not manufacturers' quotations were requested would depend on the degree of firmness of the cost estimates required and whether the design team had experience of an installation of a similar size and complexity.

(c) Detailed assessment of preferred type of machine

The third stage would involve a more complete set of process data, backed up by information on process economics and costs of start-up after an enforced

• Introduction		Part One	Chapter 1
• Identify technically feasible options		Part One	Chapter 2
• Preliminary selection of equipment type		Part One and Part Four	Chapter 2
• Detailed assessment of preferred type	Preferred type of machine		
	Fans	Part Two,	Chapter 3
	Centrifugal compressor		Chapter 4
	Axial compressor		Chapter 5
	Reciprocating compressor	Part Three	Chapter 6
	Twin screw compressors general		Chapter 7
	Oil-free twin screw compressors		Chapter 8
	Oil-injected twin screw compressor		Chapter 9
	Positive displacement blowers		Chapter 10
	Rotary sliding vane compressors		Chapter 11
• Detailed assessment of common features	Drivers and transmissions	Part Four	Chapter 12
	Lubrication		Chapter 13
	Seals for rotary machines		Chapter 14
	Inspection and testing		Chapter 15
	Containment Safety		Chapter 16
• Preparation of an enquiry sheet	Mechanical data sheet and specification		Appendix I
• Properties of gas mixtures			Appendix II

Fig 1.1   Simplified structure of the guide

shutdown (if available at this stage). This would then provide more information from which to judge reliability requirements, standby or installed spare plant and spares inventory. The more complete process data would include information on gas composition and likely particle size in the gases, whether the gases are corrosive and whether the process can tolerate oil contamination, etc.

For this third stage you should move to Part Two, Roto-Dynamic Machines or to Part Three, Positive Displacement Machines of the guide depending on whether the preferred type of machine was a fan, an axial or centrifugal compressor, a reciprocating compressor, an oil-free screw compressor, an oil-injected screw compressor, a single screw compressor, a positive displacement blower or a sliding vane compressor, and follow the procedure laid down in the relevant chapter. Part of the output from this stage would be information required elsewhere in the design of the parent plant, such as power consumption, starting loads, cooling loads, control systems, foundation requirements

ROTO-DYNAMIC MACHINES

Centrifugal fan	Axial fan	Centrifugal compressor	Axial flow compressor

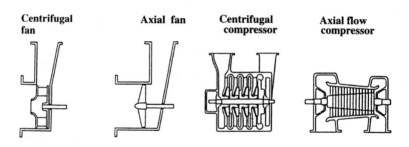

POSITIVE DISPLACEMENT MACHINES

Reciprocating compressor	Twin screw compressor	P D blower (Roots type)	Sliding vane compressor

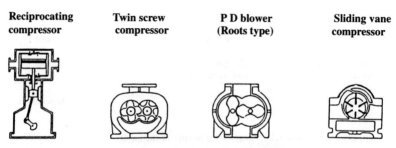

Fig 1.2   Diagrams of machine types

and size. In addition the features which are common to all of the machines would need to be considered and these are covered in Part Four.

(*d*) *Enquiry specification*

The end result of the detailed assessment from (c) above would be the preparation of an enquiry specification. This would involve the completion of a mechanical data sheet, along with the relevant machine specification (either API, ISO, EEUMA or in-house specification) to be sent out to nominated suppliers.

While it is common practice for design contractors to quote API, ISO or EEUMA Specifications without qualification it is essential that these machine specifications are read and fully understood. As they stand, without qualification, they contain hundreds of queries which need to be addressed and fully answered if purchasers wish to be assured that they are ordering a machine that will meet their full requirements.

A simplified structure of the guide is shown in diagrammatic form in Fig. 1.1, above.

A diagrammatic arrangement of the different types of fans and compressors is shown in Fig. 1.2 above.

Chapter 2

The Preliminary Choice of Fan or Compressor Type

2.1 Introduction

This chapter defines the way in which a preliminary choice of the type of fan or compressor and the number of units required may be determined from knowledge of the process duty.

The chapter includes some comments on the application of types of fan or compressor. The choice may be constrained by operating limits which are a characteristic of the machine type. Further selection constraints arise from considerations of the plant operation such as desired turndown, or acceptable level of contamination of the product by the machine itself. Some considerations on the required machine availability are also included.

Once a preliminary choice of fan or compressor type has been made, you should select the appropriate part and chapter which describes the machine type and its specific characteristics. Part Two describes roto-dynamic fans and compressors, Part Three describes positive displacement machines, and Part Four deals with items that apply to all machine types.

The selection flowsheet for preliminary choice of fan or compressor is shown in Fig. 2.1.

2.2 Establishing the duty

2.2.1 *Process data sheet*

The purpose of the process data sheet is to state the normal duty, including all operating cases, and all possible combinations of gases, pressures, flow, and temperatures, but should not at this stage contain any margins on the flowrate or pressure.

Very often the design of a machine has to be based on some duties that occur only occasionally but are essential. It is therefore vital that these duties are well defined and considered. Among the duties that should be evaluated are:

- normal process duties;
- maximum process duty;
- start up, shutdown and emergency conditions.

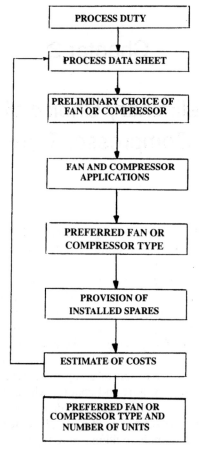

Fig 2.1   Preliminary selection sequence for fan and compressor type

Note that it is often possible to modify particularly demanding auxiliary duties in consultation with the process designers.

2.2.2   *Essential data for the completion of the process data sheet*

As a minimum the following information should be obtained:

(*a*) *Establish the normal process duty in the following terms*

– mass flows	kg/s
– gas composition	molar percentage
– absolute gas pressure	bar
– temperatures	C
– actual volume flow	m³/s

(*b*) *Gas properties at the inlet*

- gas compressibility factor, z
- ratio of specific heats
- minimum and maximum gas molecular weight attainable in any process operating condition
- corrosive constituents by quantity and type
- particle solids content by quantity, particle size and particle hardness

(*c*) *Process system requirements*
Identify any process temperature limit and establish whether oil free discharge is required from the machine.

2.2.3 Other considerations

(*a*) *Margins*
Margins on the flowrate or pressure would not normally be added when making a preliminary choice of machine, and would only be considered once a selection had been made.

Margins may be added when defining the selected machine type duty, but avoid the common pitfall of duplicating margins. Include for some or all of the following:

- uncertainties in process duty;
- uncertainties in gas properties;
- inaccuracies in performance measurement;
- machine deterioration in service.

The size of the margin added will be dependent on the machine selected, and specific values are covered in the relevant chapter.

Care should be taken with the selection of fans and centrifugal compressors where a wide range of gas densities is to be handled. For example, extreme atmospheric variations will have a significant effect on the absorbed power and driver power rating. To minimise costs the selection may be made for normal ambient conditions, accepting that in the occasional extreme condition the fan or compressor may fail to fully meet plant requirements.

(*b*) *Hazard and operability*
It is standard practice in process plant design to carry out hazard and operability studies on the plant at various stages during the design activity. An important factor in the machine selection process is ensuring containment of hazardous process fluids. The consequences of loss of containment will need to be evaluated and this may require a change in the machine selected or the addition of a specific requirement in the enquiry specification. For example, a protective trip system may be required on a containment component to prevent operation which exceeds a mechanical design limitation.

2.2.4 *The process system*

When selecting a fan or compressor it has to be borne in mind that the machine is part of a larger system into which it must be integrated. For example:

(a) Care must be taken that the gas at the first stage suction is at the design condition. A knock out vessel may be required to remove solid or liquid particulate matter which will cause erosion, and in the case of reciprocating compressors, valve damage.
(b) When machines are started up for the first time, unless the inlet lines have been thoroughly cleaned, some debris will remain in the inlet system and cause internal damage. A temporary filter may therefore be required.
(c) Intercoolers are used to limit gas temperatures. If the dew point of water or any process gas condensibles is reached it becomes necessary to follow a cooler with a separator to remove any condensed liquid before the gas enters the next stage.

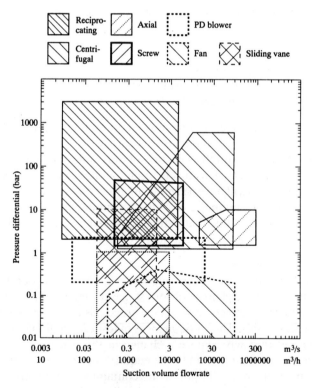

Fig 2.2 Preliminary selection diagram to determine compressor type

2.3 Preliminary choice of fan or compressor type

Operating limits for fans and compressors are summarised in Table 2.1. A fan and compressor selection chart is shown in Fig. 2.2 which provides guidance on operating ranges and areas of application.

Calculate the essential data required for preliminary selection from paragraph 2.3.1 below, and use the selection chart and the table to make the preliminary choice. You may need to refer to paragraph 2.4 which contains application guidance for fan and compressor types.

Table 2.1   Fan and compressor operating limits

The limits shown in this table are nominal for equipment which is widely available. Other physical and mechanical limits may exist which require more detailed consideration. The extreme limits shown may not be achievable for any one machine selection.

	Suction volume min m³/h	Suction volume max m³/h	Pressure ratio max	Discharge pressure max bar	Discharge temp max C	Power max kW
Fans	360	360 000			350	2000
Axial compressors	50 000	1 000 000	5 (1)	5	200	50 000
Axial/ centrifugal compressors	50 000	1 000 000	10 (1)	10	200	50 000
Centrifugal compressors	500 (2)	300 000	100 (1)	600	200	35 000
Reciprocating lubricated compressor	30	15 000	1000	3000	200	6000
Reciprocating non-lubricated compressor	30	15 000	100	100	150	2000
Diaphragm compressor	1	500	300	2500		100
Oil free screw compressors	500	20 000 (3)	5 (4)	40	225	5000
Oil injected screw compressors	50	10 000	20	45 (6)	120	5000
Positive displacement blowers	50	65 000	2	30 (5)		300
Sliding vane compressors	100	5000	10	10	180	220

Note:
(1) Dependent upon M_w of gas
(2) Discharge volume minimum 100 m³/h
(3) Up to 80,000 m³/h possible, but relatively inefficient
(4) Up to 10 possible with liquid injection
(5) Depending upon blower type
(6) 75 bar in development

2.3.1 *Essential data for preliminary selection*

The following essential data will be required to make the preliminary selection from Fig. 2.2.

(*a*) *Pressure*

Inlet pressure (absolute)	bar
Discharge pressure (absolute)	bar
Pressure differential	bar

(*b*) *Volume flowrate*

The actual volume flowrate V at the inlet can be calculated from the expression

$$V = \frac{m \cdot z_s \cdot R \cdot T_s}{100 \cdot M_w \cdot P_s} \text{ m}^3/\text{s} \tag{2.1}$$

(*c*) *Power*

Estimate the power P required from

$$P = 1.6 \, mH_{is} \text{ kW} \text{ kJ/kg} \tag{2.2}$$

where H_{is} the isothermal head, is obtained from

$$H_{is} = z_s \frac{R}{M_w} T_s \log e \frac{P_d}{P_s} \text{ kJ/kg} \tag{2.3}$$

where P_d = maximum discharge pressure bar

2.4 Fan and compressor applications

The purpose of this section is to provide guidance on the application of each type of machine. These notes should be read in conjunction with Table 2.1.

Gas density is an important property when considering roto-dynamic machine applications. Whereas positive displacement machines operate by overcoming the delivery pressure and deliver a nearly constant volume, roto-dynamic machines develop a head, and not a pressure. The pressure developed is related to the head by the density of the process fluid.

2.4.1 *Roto-dynamic machines*

(*a*) *Fans*

Centrifugal and axial fans are suitable for low pressure applications over a wide range of flowrates. Special designs are available for high temperature duties, usually above 350 C.

Although fans are normally specified with a modest aerodynamic duty, care needs to be taken with material selection and fan fabrication standards. Fans used on arduous duties (e.g. high tip speeds or contaminated fluids) will need detailed design specification.

Small fans for heating and ventilation duties are not covered by this guide.

(b) *Centrifugal compressors*

For efficiency reasons centrifugal compressors should only be specified if the discharge volume flow is greater than 0.03 actual m³/s (100 m³/h). At lower flows the efficiency will be poor. This limitation does not apply to some special compressor designs which operate with close running clearances.

As for fans, care should be taken with the selection of centrifugal compressors where a wide range of gas density is to be handled, particularly when driven by a fixed speed driver.

Discharge temperature should not normally exceed 200 C without specific analysis of compressor design and material selection. However, special compressors with a discharge temperature up to 400 C are commercially available. For some high alloy compressor casings the maximum differential temperature between inlet and discharge may be limited to 120 C.

A suction catchpot may be required to collect liquid droplets. Although centrifugal compressors are generally tolerant to atomised liquid carryover, liquid entrained with the gas can cause erosion and liquid carried over in slugs can cause severe damage.

(c) *Axial compressors*

Axial compressors are generally applied for high flow and low discharge pressures, achieving high compression efficiencies. They should only be used where the process gas is clean and dry.

Like fans and centrifugal compressors they should be avoided where a wide range of gas density is to be handled.

2.4.2 *Positive displacement machines*

(a) *Reciprocating compressors*

Reciprocating compressors are well suited to applications involving low flowrates or high pressure ratios. They can accommodate a relatively wide range of gas compositions, pressure ratios and flowrates, whilst maintaining a good efficiency. Turndown in rate can be achieved by a variety of methods (e.g. individual valve lifters, bypass of the stage or compressor). Some methods are more efficient than others.

A consideration in the selection of the reciprocating compressor is the unpredictable operation due to component failure and the requirement for routine maintenance which, in many cases, necessitates an installed spare. This requirement should be considered when comparing costs with other compressor types.

Oil lubricated reciprocating machines may subject the process gas to oil contamination. The sensitivity of the process to oil contamination needs to be checked. An important safety issue is the build up of carbonised oil in the compressor discharge system. In the presence of air this can lead to self ignition and fire or explosion. Preventive measures include selecting the correct type of oil, regular cleaning and discharge temperature protection.

Reciprocating compressors are not ideally suited to dirty or polymerising gases, and cannot tolerate liquid droplets in the suction flow without causing valve damage.

(b) Screw compressors

A screw compressor is likely to have a higher initial cost than a reciprocating compressor for the equivalent duty. However, because maintenance requirements are different, the screw compressor does not normally require an installed spare.

Dirty and polymerising gases will require careful specification and specific applications should be discussed with a manufacturer. More information on the effects of contaminated gases is given in Chapter 7.

Specific limits and applications for oil free and oil injected screw compressor types are summarised below:

(i) Oil free compressors

An oil free screw compressor is ideal for a wide variety of gases where the duty requires a low pressure ratio at constant volume flow. Capacity turndown of oil free screw compressors is inefficient, and is usually achieved by an external bypass.

Oil free screw machines are relatively inflexible and will not tolerate excessive pressure differentials or discharge temperatures.

Noise levels will be high (typically 95–105 dB(A) at 1 m).

They should be considered when liquid is entrained with the gas.

They are frequently used in vacuum applications with suction pressures as low as 0.1 bar.

(ii) Oil injected screw compressors

Oil injected screw compressors are more suitable for higher pressure ratios than oil free machines because of the cooling effect of the oil injection.

Turndown capacity can be stepless, hence they are more flexible and operate with a higher efficiency than the oil free type.

Noise levels are lower than the oil free machines (typically 85–95 dB(A) at 1 m).

Oil injected machines should be selected with care where oil carryover cannot be tolerated in the process (typically 5 ppm), or where the injected oil and process gas are incompatible.

They are generally used on applications with low molecular weight gases, most hydrocarbon gases, refrigerants, and corrosive gases.

(c) Positive displacement blowers

Positive displacement blowers are a less efficient method of compression and hence normally limited to low power, low pressure applications (e.g. pneumatic conveying of particulate material).

When fitted with shaft seals they are suitable for a variety of clean and dry gases where a low pressure ratio is required at fairly constant volume flow.

Turndown is inefficient, generally requiring bypass or blow off arrangements.

Noise levels can be high (typically 90–100 dB(A) at 1 m).

They can operate as a blower working between atmospheric pressure and above, or as an exhauster working from below atmospheric inlet to atmospheric pressure.

(d) Sliding vane compressors
Vane compressors are limited to low power, low pressure applications, traditionally in the supply of compressed air for a variety of industrial applications. Recently new specialised applications have been found in the air compression market as well as many refrigeration and process gas duties.

They generally operate at low speed and therefore noise levels are usually low. Where vanes are oil lubricated, oil carryover into the process must be considered. For continuous duties reliability may be affected by vane wear.

2.5   Provision of installed spares

Whether to install a spare machine is a key decision in the selection process, not only because of the effect on the total cost, but also because the choice of multiple machines may allow smaller machines to be used with more operating flexibility. The following factors will influence this decision:

(a) Reliability

Centrifugal and screw machines are intrinsically more reliable than reciprocating machines. For example, centrifugal and screw compressors may be specified for continuous operation for periods of up to three years. Reciprocating machines frequently require valve maintenance where the interruption to operation may be unpredictable. Although valve life can be long (typically 12 months or more on a well designed machine) valve failures do occur. A single unspared centrifugal machine may be used, whereas continuous operation on the same duty may require at least two reciprocating machines.

(b) Availability requirements

In considering the likely availability of the plant to produce its final product rather than the reliability of the compressor installation, repair time (including recommissioning time) is a significant factor. The definition of availability is

$$\text{Availability} = \frac{\text{MTTF}}{\text{MTTF} + \text{MTTR}} \qquad (2.4)$$

where MTTF = mean time to failure

MTTR = mean time to repair (including recommissioning time)

This effect of repair time on availability can often by eliminated by having installed spare machines, or mitigated by carrying major items of spares.

Table 2.2   Selection of spare machines

	Type of duty	Provision of machines
Fan	All suitable duties	1 rated at 100 percent
Centrifugal or axial compressor	All suitable duties	1 rated at 100 percent
Reciprocating compressor	Where commercial viability of process is completely dependent on continuous operation of compressor	2 rated at 100 percent or 3 rated at 50 percent. Choice dependent upon economics
Reciprocating compressor	Where plant can operate at design output with one compressor stopped	2 rated at 50 percent
Oil-injected screw compressor	All suitable duties	1 rated at 100 percent where a large enough machine is obtainable. Otherwise multiples to give 100 percent total
Oil-free screw compressor	Where commercial viability of process is completely dependent on continuous operation of compressor	1 rated at 100 percent with non-installed spare machine available in store
Oil-free screw compressor	Typically instrument or maintenance air supply or where short outages may not be very costly	1 rated at 100 percent
Positive displacement blower	All suitable duties	1 rated at 100 percent
Sliding vane compressor	All suitable duties	1 rated at 100 percent

(*c*) *Part-load performance*

The required part load performance of the plant must be considered. The costs associated with turndown efficiency, and predicted machine availability may justify the installation of multiple smaller machines. For more information on reliability and availability see reference (**10**).

As a preliminary guide to the provision of installed spares, Table 2.2 should be used for estimating the number and size of installed machines.

2.6   Estimate of costs

The total installed cost of the machine will be affected by many of the factors listed below. This list is not intended to be a detailed list of all cost factors, but only those which are likely to influence the preliminary estimate of cost.

(*a*) *Equipment costs*

– driven machine
– gearbox between machine and driver
– testing in manufacturers' works
– pulsation dampers, separators, intercoolers

- instrumentation and control system
- inlet filtration
- lubrication and seal systems
- driver systems
- baseplate

(b) *Installation costs*

- foundation
- area classification—many standard packaged fans, compressors and drivers are not suitable for installation in process plant
- noise control

(c) *Operating costs*

- efficiency at duty points
- efficiency at turndown
- maintenance requirements
- major spares

After obtaining a cost estimate which is within acceptable limits, and the preliminary choice of fan or compressor is confirmed, refer to the appropriate part and chapter to specify the principal characteristics of the chosen type. Complete the Mechanical Data Sheet as part of the enquiry preparation.

Part Two

Roto-Dynamic Machines

Foreword to Roto-Dynamic Machines

Part 2 deals with roto-dynamic machines. All these machines have in common an impeller that rotates and imparts energy to the fluid passing through it. However, the shape of this impeller may vary from the shape of that of a little pump to a shape similar to the sails of a windmill.

Dimensionless numbers

To deal with the diverse shapes of impellers, all operating in a similar way, various dimensionless numbers are in common use. The most widely used ones are the:

pressure coefficient	ψ,
flow coefficient	φ,
diameter coefficient	δ and
speed coefficient	σ

Any pair define a given impeller shape. The first pair is easily understood:

The pressure coefficient ψ is the ratio of the achieved pressure compared with the kinematic pressure obtained using the tip speed in Euler's formula, i.e.

$$\psi = 2H/u^2 = 2\Delta p/\rho u^2 \qquad (A2.1)$$

It is of the order of one for centrifugal machines though less for axial machines. Note that it can be greater than one due to the centrifugal effects in the impeller.

The flow coefficient

$$\varphi = Q/(uA) \qquad (A2.2)$$

The area A can refer to various areas of the compressor: the area of the impeller, $(\pi/_4D^2)$ or the exit area, (πDb), where b is the width of the outlet of the impeller, are the most common ones used for centrifugal machines. For axial machines the area of the annulus between rotor and stator is frequently used. Care must be taken to specify which area is used before making any comparisons between data from different authors. The flow coefficient, if based on the exit area, can loosely be compared with the volumetric efficiency of positive displacement machines. The above two dimensionless numbers are mainly used when dealing with compressors.

For fans the second pair, the Diameter and the Speed Coefficient are particularly useful. They are defined as follows:

Diameter Coefficient $\delta = 1.05D(\Delta p/\rho Q^2)^{1/4}$ (A2.3)

and

Speed Coefficient $\sigma = 2.108\,N(Q^2\rho^3/\Delta p^3)^{1/4}$ (A2.4)

The two pairs are related as follows:

$$\psi = 1/\sigma^2\delta^2 \tag{A2.5}$$

$$\varphi = 1/\sigma\delta^3 \tag{A2.6}$$

and

$$\delta = \sqrt[4]{(\psi/\varphi^2)} \tag{A2.7}$$

$$\sigma = \sqrt[4]{(\varphi^2/\psi^3)} \tag{A2.8}$$

The use of the above coefficients allows general data to be used in the selection and in the design of roto-dynamic machines.

Head

A major difference between positive displacement compressors and dynamic machines is that the former raise the potential energy of the fluid by reducing its volume and so increasing its pressure. The pressure is raised till it exceeds the pressure at the delivery. In the first approximation it does not depend on the flowrate which is constant.

Dynamic machines—axial and centrifugal pumps, fans and compressors—work mainly by accelerating the fluid handled, so increasing its velocity. This leads to an increase in kinetic energy inside the moving parts (the impeller) which is then converted to potential energy—'head'—in the stationary parts. The 'head' available at the delivery varies significantly with the flow through the machine. 'Head' is the energy added to the fluid. It can also be expressed as the height of a fluid column to which the handled fluid can be raised. The pressure at the bottom of such a column of fluid is proportional to the density of the fluid. For example, a pump used on mercury and water will lift both to the same level, but the pressure at the pump outlet will be 13 times greater if mercury is pumped rather than water.

Head for a liquid can be physically measured: it is the column of liquid, e.g. water. The head of a gas column is more difficult to establish as the gas changes its density in the column. It is denser at the bottom than at the top. The head can be expressed in various ways. In Part two of this guide the polytropic head and efficiency are always used. These two quantities characterise the performance of a stage in a roto-dynamic machine.

The polytropic head H can be calculated from the gas conditions and the pressure ratio Π

$$H = R/M_W T_S zn/(n-1)(\Pi^{(n-1)/n} - 1) \quad \text{kJ/kg} \tag{A2.9}$$

Note that the value of the polytropic coefficient has little influence on the polytropic head, it does however significantly affect the polytropic efficiency. The polytropic efficiency η_p is defined as

$$\eta_p = (1 - \kappa)n/\kappa(n - 1) \tag{A2.10}$$

If the polytropic efficiency is given the polytropic index can be calculated from

$$n = \eta\kappa/\{1 - \kappa(1 - \eta)\} \tag{A2.11}$$

Note that on existing machines the polytropic efficiency η can be calculated from knowledge of inlet and outlet conditions of a stage.

$$(n - 1)/n = \lg(T_2/T_1)/\lg\Pi \tag{A2.12}$$

and therefore η

$$\eta = (\lg\Pi/\lg(T_2/T_1)) (1 - \kappa)/\kappa \tag{A2.13}$$

Power

The gas power required per stage is

$$P = mH/\eta_p \tag{A2.14}$$

A stage is defined as a part of a machine where energy is added but not removed, i.e. no cooler.

For fans with a low head the formula simplifies to

$$P = 100 \, Q\Delta p/\eta_p \tag{A2.15}$$

Mechanical losses have to be added to the power needed to compress the gas.

Chapter 3

Fans

3.1 Introduction

It is assumed that you have already been through the process of preliminary machine selection (Chapter 2) and that a fan is appropriate for the duty required. It must be realised that the word fans covers a variety of different machines of different design and that this chapter covers a larger range of machinery than any other in this guide. This chapter provides a method for finding a suitable fan unit based on flowrate, pressure rise, density and type of fan favoured.

Alternatively if the shaft speed is one of the chosen starting parameters a type will be recommended. Some guidance on the suitability of the design is also given. However, the guide does not give sufficient detail to allow a fan to be designed.

Flow through a fan is complicated. Quoting from Eck (**11**): 'Owing to the considerable uncertainty of the behaviour of the boundary layer in the rotating passage, the theoretical design of fans appears to be impossible without exhaustive test bed measurements. Therefore good testing is essential in all aspects of serious work with fans.'

The dimensions and characteristic obtained using the method outlined below can be used to obtain a specification for enquiry purposes and to provide preliminary information for design work by others. It should, however, be noted that fan design is not standardised and manufacturers have their own bespoke design methods. It is therefore strongly advisable to confirm a choice with a selected vendor before basing a layout on the data obtained from this chapter of the guide.

Although the traditional terms of fan, blower and compressor are extensively used, they have no universally accepted definitions. In this guide a fan is arbitrarily defined by a duty envelope appropriate to the safe use of simple design methods and constructions, where low production cost is the dominant consideration. Small fans for heating and ventilating duties, cross flow fans and large propeller fans are not covered by this guide. Figure 3.1 gives the sequence of the decision making process for the selection of centrifugal and axial fans.

Note that before a selection can be made the design point or points for the fan must be specified. These points are found by adding all necessary margins to the rate and head required by the process design.

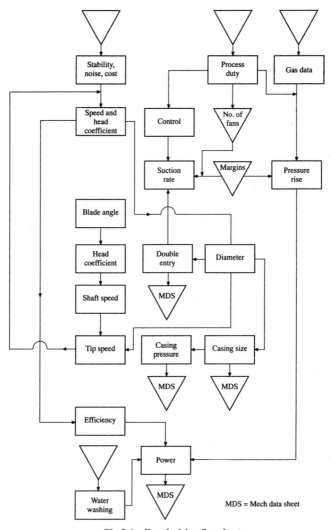

Fig 3.1   Fan decision flowsheet

3.2   Basic data

3.2.1   *Pressure rise*

When dealing with dynamic compressors it is usual to use 'head' and 'flowrate' to define an operating point. Fans too are dynamic machines and the same nomenclature could be used. However, with fans it has been practice to express the head in terms of the pressure rise referred to an inlet density and temperature of 1.2 kg/m^3 and 20 C respectively. Catalogues are based on this, being the industrial standard, but it has the disadvantage that if a fan has to operate at different temperatures a different set of curves, or at least a separate scale, is needed for each temperature or other change in density.

The energy added by a fan to the gas appears partly as potential energy (head or pressure rise) and partly as kinetic energy (velocity). Some fans provide mainly kinetic energy for example cooling fans. They impart a velocity to the air, or gas handled with a very small or even no pressure rise. Other fans are designed to raise the pressure of the medium handled. In that case it is necessary to convert the kinetic energy remaining at the exit of the impeller into potential energy, rather than lose it all in turbulence and friction. To do this a diffuser or volute is needed. After such a device some of the energy added by the impeller is still present as kinetic energy, i.e., velocity. This remaining energy is present at the outlet of all roto-dynamic machines. It does not contribute to the available head and, as it is of the order of 5 mbar, can not be ignored when dealing with fans, though with machines with large pressure rises it is not significant. With fans, however, it represents a significant part of the available energy and must therefore be considered.

The definitions for the pressure rise in a fan used in the British Standard for Fan Testing (BS 848 para 9) (**12**) vary with the fan ducting, as follows:

(a) Fan installations are defined as A, B, C or D depending on the ducting. A is without inlet or delivery duct and could be diaphragm mounted. B and C are with an outlet and inlet duct respectively and D with both inlet and outlet duct.

(b) For types A and C the pressure rise is the difference between the static pressure at the delivery and the total pressure at the inlet. To obtain the total pressure add the dynamic pressure at the delivery to the static pressure.

(c) For types B and D the fan total pressure is the difference between the total pressure at delivery and inlet; to obtain the fan static pressure subtract the fan dynamic pressure at the delivery.

In this guide the design calculation and efficiencies are based on total pressures, i.e. the sum of static and kinetic pressure. To calculate the duty required from fans of type A and C, a pressure equal to the head of the exit velocity (the fan dynamic pressure at the delivery) must be added to the head specified to fulfil the process duty. Increase the head by 5 percent to allow for the dynamic head at the outlet when estimating fan size. When comparing designs offered a more rigorous approach must be followed, based at the velocity at the outlet, to ensure that we compare like with like. The use of 'Total Pressure' as used in this guide facilitates the design process. However, users need to know the static pressure rise to compare machines offered for a given duty and for this reason the remaining kinetic head must be known.

Process data commonly specifies the pressure required for the process—the fan will not only have to supply that pressure but also the pressure needed to overcome the friction losses in any ducts leading to and from the fan.

3.2.2 Flowrate

Flowrate is usually specified as a dry mass flow in kg/s. The design of machinery however is based on actual inlet volume. The required mass rate must be expressed as a volume flow at inlet conditions. To achieve the required mass flow under all conditions the design rate and head of the machine must be based on the conditions with the lowest density and pressure at the inlet. (Note that inlet density is affected by pressure, temperature and humidity. See Appendix II on how to calculate the density.)

3.3 Margins

Once the design rate and pressure rise have been settled margins should be considered for the following factors:

(a) Tolerances

Most specifications allow a plus/minus tolerance so that the fan is said to meet its duty even if the achieved performance is below the specified design point. If it is mandatory to achieve the design point add the negative tolerance, 2.5, 5, and 7.5 percent for fans tested to BS 848 Class A, B and C respectively. Note that the German code (**13**) does not permit negative tolerances.

(b) Uprating

To allow for future uprating, which is particularly important in mine ventilation fans where the required rate and head increases with increasing underground workings.

(c) Deterioration in service

To be able to maintain rate and head in case of a deterioration due to build-ups or erosion.

(d) Control

All control devices cause a pressure drop. This must be allowed for.

(e) Margins customary in the industry where the fan is to be used

These should be carefully examined, and their justification proven and checked so that they do not duplicate the margins mentioned above.

Fans are usually used in systems without a static rise but where the required head is proportional to the square of the flowrate. A need to be able to increase the flow by X percent will therefore require the head to be increased by 2X percent and the power by 3X percent. Adding large margins means that operation at the likely operating point is at a significantly reduced efficiency. Always ensure that all margins that are added to the design requirements are valid and kept to a minimum. It has already been stated above that being over-generous with design margins will provide an oversize fan with the normal operating duty on a low efficiency point on the characteristic field in addition to the losses incurred when a damper is used to control the flow.

3.4 Design point

By adding the margins to the specified process duty the fan duty point is arrived at. Note that fan manufacturers are normally willing to guarantee only one duty point. However, in an enquiry include all points with limiting conditions, i.e.:

(a) maximum capacity Q_{max} with the corresponding pressure rise or head specified as a 'minimum' value that must be achieved
(b) maximum pressure rise Δp_{max} with the rate given as a minimum value that has to be achieved

3.5 Factors affecting centrifugal fan selection

Figure 3.2, besides giving a relation between speed and diameter coefficients (curve 'δ'), and the maximum possible efficiency (curve 'η'), shows a series of fan shapes. These shapes are chosen such that each one will have the same design point, i.e. rate and head, provided the tip speed u is increased with increasing speed coefficient as shown in the third curve 'u'. In theory any type of fan design (i.e. radial or axial), can fulfil any duty. In practice, however, there are

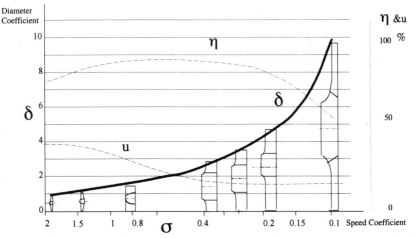

Fig 3.2   Speed–diameter coefficients. (Source: Eck, B. *Fans – Design and operation of centrifugal, axial and crossflow fan*, Pergamon Press)

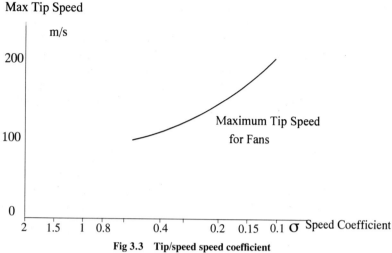

Max Tip Speed

Fig 3.3   Tip/speed speed coefficient

many factors that limit and influence the choice. Among these tip speed is the most important restriction.

The highest recommended tip speeds for fans are shown in Fig. 3.3. Note these maximum speeds apply only to fans; if they are exceeded the machine no longer conforms with the definition of a fan used in this guide but falls under the category of compressors. At very low speed coefficients 190 m/s is the highest tip speed at which a machine can still be considered a fan. However, it must be remembered that this limit applies only to fans with large diameter coefficients and low speed coefficients. The maximum allowable tip speed for fans with smaller diameter coefficients is much lower as indicated in Fig. 3.3. The speed is limited by the fan stresses which, for a given shape, are proportional to the square of the tip speed. At elevated temperatures the allowable tip speed is less due to the reduction in proof stress of the impeller material with increasing temperature.

The highest heads can only be obtained with impellers having low speed coefficients, but for pressure rises between 1000 and 10,000 N/m² there is a choice of types with different speed coefficient. The cheapest design is usually the fastest fan but considerations other than first cost may govern the selection. Some guidance is given below.

3.5.1   *Fan type*

Some of the factors that influence the type of fan selected are:

(*a*)   *Cost*
Is capital cost the main concern? If so select the design with the highest rotational speed N that meets the head and flow requirement.

(b) *Tip speed*
Stresses in an impeller of a given shape are directly proportional to the square of the tip speed. For corrosive or hot operating conditions do not select a fan having maximum tip speed.

(c) *Drive speed*
Most fans are motor driven. Using a direct drive saves the cost of a gearbox. Further 4, 6 and 8 pole motors are the most price worthy (i.e. cheapest) drivers, their use is therefore recommended, giving, in Europe, speeds of 25, 16 and 12.5 Hz (1500, 1000 and 750 rpm) respectively.

(d) *Efficiency*
Figure 3.2 shows that peak efficiencies are obtained if the speed coefficient σ is between 0.3 and 0.8. Note that the efficiency of centrifugal fans increases with increasing hub ratio $D_r(d/D)$.

(e) *Noise generation*
This increases with the 5th power of tip speed and noise generation is lowest with a low tip speed and a large pressure coefficient ψ.

(f) *Wear due to dust*
Again a large pressure coefficient ψ is required. Radial blades are preferred.

(g) *Capacity for a given fan size*
A maximum value of the flow coefficient φ is required.

(h) *Minimum size and cost for a given performance*
The product of pressure and flow coefficient ($\psi\varphi$) should be a maximum. This is achieved with centrifugal fans having forward sloping blades and large inlet areas, i.e. inlet diameters close to those of the outside diameter. Drawbacks of these designs are their low efficiency and 'wavy' characteristic which makes control difficult and operation in parallel impossible.

To obtain further details see 'Fans' by Eck (**11**).

3.5.2 *Stability*

An absolutely stable characteristic is defined as one where the fan pressure decreases continuously with increasing flow over the full operating range from zero to maximum flow. Efficient commercial fans rarely have this characteristic, but it is inherent in open impellers, i.e. the paddle type.

'Conditional' stability is sufficient for most duties and should be specified at the enquiry. It is defined by a characteristic where in the operating region the head decreases with increasing flow (see Fig. 3.4) and:

(a) there is an increase of at least 15 percent in the fan pressure rise Δp from the rated point to the highest point on the curve
(b) the highest point on the curve occurs at a flow less than 90 percent of any

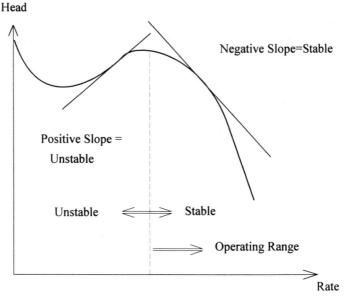

Fig 3.4   Conditional stability

predicted process minimum flow requirement, and not more than 50 percent of the flow at the best efficiency point

(c) no system characteristic intersects the $V - \Delta p_s$ characteristic at more than one point

3.5.3 *Parallel operation*

When two or more fans are to run in parallel, it must be specified:

(a) that the duty range lies not only in the stable part of any one fan but also within the stable part of the combined characteristic, and

(b) that all fans shall have conditional stability and a non overloading power characteristic with their drivers rated for the maximum power the fan can absorb

3.5.4 *Stresses*

It has been stated above that tip speed is normally the limiting parameter in designing a fan. The tip speed in turn is limited by the stresses set up due to centrifugal forces. The wider the exit of the impeller, i.e. the greater the ratio of b/D, the higher the stresses and the lower the permissible tip speed. Calculating the stresses set up in impellers is beyond the scope of the guide; suffice to say that the lower the speed coefficient, the lower the stresses at a given tip speed.

3.5.5 Blade angle

The shape of the characteristic of fans is mainly dependent on the exit blade angle, see Fig. 3.5. This can be radial −90 degrees, forward sloping <90 degrees and >90 degrees backward sloping. The pressure coefficients are largest for forward sloping blades, up to 3; between 1.2 and 2 for radial blades depending on the speed coefficient, increasing with rising speed coefficient. They are lowest for backward sloping blades, about 0.8 or less. Fans with backward sloping blades are the ones most commonly used in industry due to their high efficiency and the stability of their operating characteristic. Figure 3.6 shows the operating characteristics of the various designs.

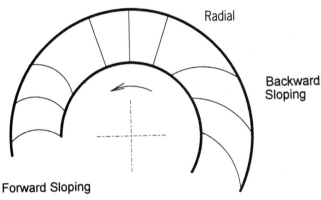

Fig 3.5   Blade angles

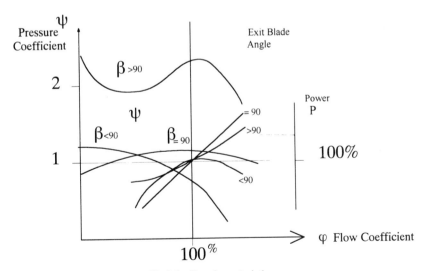

Fig 3.6   Fan characteristics

3.5.6   *Blade shape for fans with backward sloping blades*

The optimum blade type and shape is influenced by the rate and contamination of the gas handled. For clean gases with liquid and dust (the dust must be non sticking and its particle size predominantly between 20 and 200 μm) loadings w_l and w_s of less than 40 and 0.1 gm/m^3, respectively.

Flat backward sloping blades should be chosen if

$$Q_s \Delta p < 40\,000. \tag{3.1}$$

For larger duties i.e.

$$Q_s \Delta p > 40\,000 \tag{3.2}$$

Aerofoil as well as bent flat plates offer a better efficiency. If, however, the solid content is erosive and above 0.1 gm/m^3 flat plates are preferred.

For very dirty and wet applications (liquid and dust content above 40 and 1 gm/m^3 respectively) radial (paddle) blades are most suitable.

3.5.7   *Single versus double entry fans*

Most fans are single entry with an overhung impeller. This leaves the eye of the impeller completely clear and results in a high efficiency. Double entry fans are supported on both sides with a shaft running through the eye of the impellers. This leads to a slightly lower efficiency and requires seals more sophisticated than those needed for overhung impellers. Further the duct work needed at the inlet is much more involved than that for a single entry fan. However, for a given duty they are of smaller diameter and run faster and, particularly if they have atmospheric suction so that no inlet ducting is required, they are cheaper. In addition they operate with a lower noise level if the blades in the two halves are offset. To use the calculation method outlined below enter half the actual flowrate for double entry fans.

3.6   Axial fans

There are three main categories of axial fans:

(a) Free fans. This type rotates in a free space. They are used to circulate air.
(b) Diaphragm mounted fans.
(c) Ducted fans.

Free fans are only used for ventilation duties. Diaphragm fans are used to circulate air; they are only capable of giving very low pressure rises. These guidelines deal only briefly with diaphragm and ducted fans. Note that diaphragm mounted fans are increasingly replaced by fans with short ducts due to the latter's higher efficiency and lower noise level.

Allowable tip speeds of axial fans are lower than those of radial fans. This, together with lower pressure coefficients, limits the attainable pressure rise.

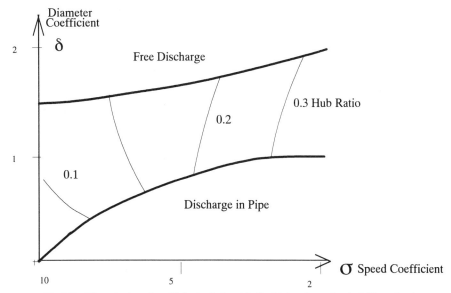

Fig 3.7   Diameter/speed coefficient relationship for high values of σ (axial impellers)

Machines with high tip speeds and multistage machines are no longer of simple construction and are dealt with in Chapter 5: Axial Compressors.

An extra parameter needed for the design of axial fans is the ratio of hub diameter to outside diameter $Y_r(d/D)$. The smaller this ratio the higher the speed coefficient, propeller fans having the highest.

Figure 3.7 is a more detailed representation of speed and diameter coefficient for axial fans. On this figure there are two parameters. The first relates to the minimum hub ratio that can be selected for a given speed or diameter coefficient, the second to the type of installation. The lower line represents a fan installed in a duct, the upper one a fan with a free discharge. Note that the method recommended in this chapter ceases to give any sensible design if σ is greater than 2 due to the added complexity of such fans. The method should therefore not be used to obtain a preliminary design.

To obtain the high efficiencies possible with axial fans they have to be precisely machined as well as have a polished surface finish on the impeller blades. This tends to increase the cost significantly. Wood was, and still is, a material often used for propeller fans—it is easily machined and takes a high surface finish.

In axial fans a rotating movement is given to the gas. On high efficiency fans this is straightened by either having guide blades (a stator) in front or after the impeller or by having a counter rotating impeller.

While it is easy to specify the rate required for a given duty, e.g. cooling air, it is much more difficult to estimate pressure drop of a system accurately. Axial impellers have a very narrow working range, see Fig. 3.8, and to allow the fan

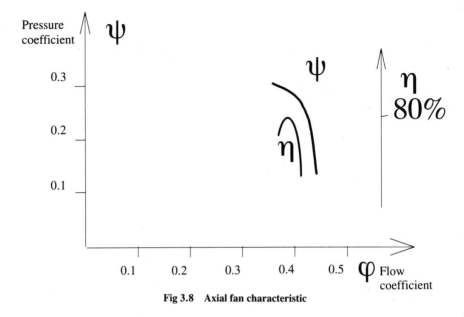

Fig 3.8   Axial fan characteristic

to be matched to the system, rotor or stator blades are often arranged so that their angle of attack can be adjusted. The calculation of blade angle and twist is beyond the scope of this guide. The number of blades should be such that the distance between the tips of adjacent blades is of the same order as the length of the blades; i.e. the smaller the hub to tip diameter ratio the fewer blades there should be.

3.7   Control

3.7.1   *General*

Control over the rate is frequently required. Various devices are available and their application and operation is described below.

It should be noted that many of the more elaborate, fast acting control systems specified at the design stage are frequently abandoned in practice and a fixed operating characteristic is accepted, though the facility of being able to select an operating point in the first place, using the control device, is most desirable.

3.7.2 *Damper*

This should be first choice for smaller fans if the dust load is non sticky and low,

$$Q^* \Delta p < 100\,000 \tag{3.3}$$

and $w_s < 5$ (w_s is the suspended solids content of the gas, in gm/m^3)

A slight power saving occurs with dampers installed on the inlet side. However, any damper must be far away from the fan inlet so as not to disturb the inlet flow pattern. This is particularly important with double entry fans where a non-symmetric flow can lead not only to one half of the fan surging but also to excessive axial loads. For this reason dampers are usually installed downstream of double entry fans.

Dampers are also used to supplement other control devices as only dampers allow operation at very much reduced rates (less than 33 percent of design).

Even in the fully open position, dampers will cause some pressure drop. When lacking experimental data you should increase the design pressure by one or two velocity heads (at the damper) to allow for the pressure drop across the damper in the open position.

3.7.3 Inlet guide vanes

Inlet guide vanes allow flow control by inducing a swirl in the gas before it enters the impeller and so changes the rate without lowering the efficiency. They are, therefore, used on large machines, though they act as a damper when approaching their closed position. They should be used, provided the gas is clean and dry, i.e.

$$w_s < 0.1 \ w_l < 10 \quad \text{and} \quad Q\Delta p > 100\,000$$

where w_l the suspended liquid content of the gas in gm/m^3.

3.7.4 Variable speed

Variable speed should be selected where large variation in inlet conditions and gas composition have to be accommodated.

It should also be used where there is a heavy solid contamination. In particular wear due to abrasive particles increases rapidly with speed. Speed control allows the operating speed, and therefore wear, to be kept at a minimum. Figure 3.9 shows how speed affects the characteristic of a centrifugal fan. The shape of the characteristic is typical of a centrifugal fan.

3.7.5 Multiple fan installation

Two or more damper controlled fans operating in parallel but arranged for sequential autostop and autostart action should be considered if large variations of flowrate at constant pressure are the anticipated flow pattern. See paragraph 3.5.3 for design requirements.

3.8 Selecting a fan

To select a fan proceed as follows.
   Calculate:

(a) The density ρ of the gas handled. See Appendix II.

(b) The pressure rise (Head) Δp. See paragraph 3.2.1. Do not omit to include duct pressure losses on the inlet and exit as well as the kinematic head at the outlet of the fan.

(c) The required flowrate Q in m³/s. See paragraph 3.2.2. Have all margins been included?

(d) Select type of fan favoured (sketches on Fig. 3.2). Read off speed coefficient σ.

(e) Find the diameter coefficient δ for a best efficiency fit from Figs 3.2 and 3.8 for centrifugal and axial fans respectively.

(f) Calculate the shaft speed:

$$N = 0.5\ \sigma(\pi Q)^{-1/2}(2\Delta p/\rho)^{3/4} \tag{3.4}$$

If close to a speed obtainable from a direct drive alter the speed coefficient to obtain such a speed.

(g) Calculate diameter, tip speed, flow and pressure coefficient:

$$d = 0.94\ (\rho Q^2/\Delta p)^{1/4}\delta \tag{3.5}$$

$$u = \pi dN \tag{3.6}$$

$$\psi = 1/(\sigma\delta)^2 \tag{3.7}$$

$$\varphi = 1/\sigma\delta^3 \tag{3.8}$$

(h) Check that the tip speed is less than the maximum permissible for the type of impeller chosen and that the pressure coefficient is acceptable.

The above calculations are best done on a spreadsheet (see Table 3.1).

(i) If the tip speed is not within the permissible range: if it is too high you should lower the speed coefficient.

Repeat d–f.

If too low is a more economic fan wanted? If so increase speed coefficient and repeat d–f.

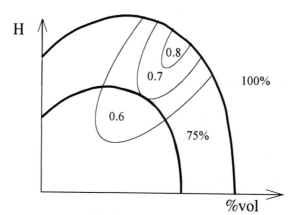

Fig 3.9   Speed variation

(j) Check if pressure and flow coefficient are acceptable.
The pressure coefficient depends on the blade angle. As stated above most industrial centrifugal fans have backward sloping blades with a pressure coefficient less than one.

The procedure outlined above can be used for axial fans. However, particularly with smaller fans, where it is customary to drive such machines directly, more emphasis should be placed on direct drives.

To use drive speed as one of the given parameters, together with rate and head, calculate the speed coefficient σ using equation (A2.4) and choose δ from Fig. 3.2 for maximum efficiency. The impeller diameter D can then be estimated using equation (3.5) and the tip speed from (3.6). Check if the impeller shape and tip speed are suitable.

3.9 Power

Use the power rating from the following calculations only for preliminary estimates. The calculation outlined below assumes the fluid handled to be incompressible. This is justified provided the pressure ratio $(p_s + \Delta p)/p_s$ is small <1.25 and adequate for a preliminary design.

For larger pressure ratios use the formula given in the Foreword of Part 2, equation (2.14).

Check the actual power rating required after selecting the fan supplier.

(*a*) *Efficiency*

Take the efficiency from Fig. 3.2. This is the peak efficiency obtainable for large impellers ($d > 0.5$ m) of first class design. This peak efficiency applies only at one point. As one moves away from this point the efficiency decreases due to increasing shock, entry, and diffuser losses (see Fig. 3.7). Further the maximum obtainable efficiencies given in Fig. 3.2 apply only for impellers with backward sloping blades. They are lower for impellers with radial and forward sloping blades. For these assume an efficiency 40 percent lower than the highest possible efficiency i.e. $\eta = 0.6\eta_{max}$.

(*b*) *Calculation of absorbed power*

$$P = mH/\eta \tag{3.9}$$

$$\text{or} = Q100\,\Delta p/\eta \text{ in kW} \tag{3.9a}$$

(*c*) *Special considerations*

When calculating the absorbed power the following should be considered:

(i) For wet gas applications the motor should be sized for dry gas at the minimum cold start temperature.

(ii) With very high dust loads allow for the increase in gas density,

Power corrected $P_c = P(\rho + w_s)/\rho$ (3.10)

(iii) For hot gas fans required to start in a cold condition check that the rating chosen is suitable for a cold start with maximum air density.

    If capacity control is provided the power required for a cold start can be reduced by throttling the flow, except if the process requires full mass flow even with cold air.

(iv) Fans with backward swept blades have non overloading power characteristics, see Fig. 3.6. Vendors should be requested to state peak power demands in addition to the power required at the duty point.

(*d*) *Radial and forward bladed*

These fans have an overloading power characteristic, see Fig. 3.6. Motor power to be specified so that at maximum flow (stonewalling) the motor is not overloaded.

(*e*) *Margins on driver power*

These fall into two categories:

(i) those always applied (10 percent?)
(ii) those added to enable future uprating

(*f*) *Water washing*

This increases the power demand. Allow 5 percent margin to cover for this.

3.10 Drivers

Electric motors are the most common driver used with fans. Variable speed drives are occasionally specified, a single speed or sometimes a two speed motor is more common. Two speed motors are particularly favoured if future uprating of the fan duty is envisaged. Small fans are usually either directly or 'V' belt driven (up to 200 kW). Larger fans are driven either directly or through a gear box.

    Fans have large moments of inertia. This often leads to problems when electric motors are used with direct on line start. Double entry fans have a lower moment of inertia than a single entry fan for a similar duty.

3.11 Casing

3.11.1 *Size*

The approximate diameter of the casing of a centrifugal fan is twice the size of the impeller diameter. The length, including the motor for direct driven machines is about 3.5 and 5.5 times the diameter of the impeller for single and double flow fans respectively.

3.11.2 Pressure rating

The casing must be able to withstand the highest pressure that can occur inside the fan. This pressure is based on the sum of the highest likely barometric pressure, or the relief valve pressure in the case of an inlet vessel, and the highest pressure the fan can generate. The latter is about 115 percent of the static pressure of the fan adjusted for the increased inlet pressure. For small fans the strength of the casing is not based on this pressure but on the need to avoid drumming and flexing as well so as to limit noise transmission.

Maximum pressure

$$p_{max} = p_{bmax} + p_{smax} + \Delta p_{max} \qquad (3.11)$$

Note that p_{max} and p_{bmax} (barometric pressure) are absolute pressures. p_s is a gauge pressure in this formula!

3.11.3 Test pressure

The casing shall be tested at 150 percent of the pressure above the barometric pressure

$$p_{test} = 1.5(p_{max} - 1) \qquad (3.12)$$

P_{test} is a gauge pressure

3.11.4 Rating for explosion

For processes handling explosive dust concentrations, specify the design to an appropriate design code (e.g. National Fire Protection Association (USA) Code 91) (14).

Obtain a value for the maximum pressure that can be generated in such an explosion. Specify the casing to withstand that pressure. Test pressure shall be 200 percent of design pressure.

Where escape of hot gases may cause a hazard external to the fan casing, mechanical shaft seals or other approved positive seals should be specified.

Rating for detonation lies outside the scope of this guide.

3.12 Ducting

Fans directly discharging into a plenum chamber may create process problems due to non uniform distribution of flow to the process especially when the speed coefficient is above 0.5. If such problems are likely the length of the discharge duct, including any in-line silencer, should be longer than 7 × (duct area)$^{1/2}$. If this is not practical then it should be specified that the fan has a uniform distribution of the discharge flow. This may be aided by locating a multi-vane damper at the fan discharge.

3.12.1 *Inlet duct layout*

Fans with ducted inlets should have a minimum length of straight duct of

$$1 = 3(\text{duct cross-sectional area})^{1/2} \tag{3.13}$$

If this straight length can not be accommodated, investigate the use of flow straighteners. These may take the form of vaned bends.

Fans with an open inlet should be provided with a bell mouth entry and, for fans with a low pressure rise (<600 N/m^2) a gust shield should be provided. This will lessen, but not eliminate, transient flow variations.

3.13 Materials

Generally the materials of construction used on the fan casing and impeller are carbon steel. Grades of steel for the impeller will vary depending on the temperature of operation and the tensile strength requirements. Where the process gas is of a high temperature (greater than 300 C), and/or corrosive or erosive, special materials are used such as stainless steel and a range of nickel alloys—though the use of exotic materials means that the design and fabrication are no longer simple. Simple design and construction is one of the conditions given above in the definition of a fan.

The impeller hub is normally of cast steel or fabricated from carbon steel. Cast iron is also used, specially for lightly loaded fans.

In wet gas applications, where acid dew point corrosion presents a problem, the casing and occasionally the impeller, can be lined with a protective layer, usually a type of rubber. If the impeller is to be lined the tip speed shall not exceed 130 m/s for radial and backward inclined blades. For other types consult the vendors about their proven experience. In most applications a stainless steel impeller is preferable.

In all cases the impeller should be of welded construction, not riveted, with the hub bolted or welded on to it. Shafts should be made from one piece of material, an exception are shafts with a heat break. Shafts are normally forged or made of bar stock.

3.14 Noise

Fans, particularly large ones, are major sources of noise. Measures to decrease noise, once a fan is installed, are difficult and extremely costly. It is, therefore, necessary to specify permissible noise output at an early stage during the design process so that if required suitable silencers can be included. Note that silencers have a length to diameter ratio of about 3.

3.15 Example

On a fertilizer plant it is necessary to select an air fan for the following duty:

Flowrate	$Q = 200 \text{ m}^3/\text{s}$ actual
Suction pressure	$p_s = 90 \text{ kN/m}^2$ abs (0.9 bar)
Pressure rise	$p = 7600 \text{ N/m}^2$
Suction temperature	$T_s = 30 \text{ C}$
Density $= \rho_0 p_1 T_n/(p_n T_i)$	

$$= 1.29 \times 0.9 \times 273/(1.02 \times 303) = 1.026 \text{ kg/m}^3$$

The 'equivalent' pressure rise: $7600 \times 1.20/1.026 = 8890$ Pa

Choose speed coefficients between 0.38 and 0.45. Select the corresponding diameter coefficients from Fig. 3.2. Using equations (3.4) to (3.8) in a spreadsheet, see Table 3.1, gives a selection of designs with an efficiency of about 85 percent. Select the design with the speed coefficient of 0.44. Those with higher speed coefficients have excessive tip speeds.

Shaft speed of 11.68 rps, diameter of 3.49 m, tip speed of 127 m/s,
flow coefficient of 0.164 and pressure coefficient of 0.9,
The shaft power would be $200 \times 7.6/0.85 = 1800$ kW.

The fan would have a diameter of 7 m and the length of the fan and motor would be 12 m. The pressure coefficient of 0.9 allows the use of a fan with backward sloping blades. If a directly coupled 8 pole induction motor is chosen the tip speed would be slightly in excess of those recommended in Fig. 3.2. This

Table 3.1 Selecting a Fan with the Help of a Spreadsheet

1 Rate	Q	m^3/s					
2 Head	Δp	N/m^2		Given data			
3 Density	ρ	kg/m^3					
4 Speed coefficient	σ			Chosen from			
5 Diameter coefficient	δ			Fig 3.2 for optimum efficiency			
6 Speed	N	r/s	$0.47\sigma Q^{-0.5}(\Delta p/\rho)^{0.75}$				
7 Diameter	D	m	$0.95\delta(\rho Q^2/\Delta p)1^{0.25}$				
8 Tip speed	u	m/s	$\pi N D$				
9 Flow coefficient	φ		$1/(\sigma\delta^3)$				
10 Pressure coefficient	ψ		$1/(\sigma^2\delta^2)$				
Flowrate	Q		200.00 m^3/s				
Pressure rise	Δp		7600.00 N/m^3				
Density	ρ		1.026				
Speed coefficient	σ	0.42	0.43	0.44	0.45	0.46	0.47
Diameter coefficient	δ	2.50	2.45	2.40	2.35	2.30	2.25
Shaft speed	N	11.15	11.41	11.68	11.94	12.21	12.47
Diameter	D	3.62	3.55	3.48	3.40	3.33	3.26
Tip speed	u	126.76	127.19	127.49	127.67	127.73	127.67
Flow coefficient	φ	0.15	0.16	0.16	0.17	0.18	0.19
Pressure coefficient	ψ	0.91	0.90	0.90	0.89	0.89	0.89

results in high impeller stresses, hence particular emphasis should be placed on manufacture and inspection.

A double flow machine for the same duty would have an impeller diameter of 2.5 m and could be driven by a directly coupled 6 pole motor. The casing diameter would be about 5 m and the total length of fan and motor would also be 12 m.

Chapter 4

Centrifugal Compressors

4.1 Introduction

It is assumed that you have already been through the process of preliminary machine selection (Chapter 2) and that a centrifugal compressor is appropriate for the duty required. Centrifugal compressors are of the roto-dynamic type which function by generating head.

The head generated is fixed by the impeller dimensions, particularly by the blade exit angle, the tangential tip speed and slip. The head seen at the delivery, because of internal losses, depends also on the flowrate. However, the head required is determined by the process conditions.

It is particularly important to identify all of the process duties to be met by a centrifugal compressor. Once the compressor dimensions are fixed, duties other than the specified range may not fall within the compressor operating envelope or can only be accommodated, if at all, by inefficient operation.

4.2 Characteristic

A typical centrifugal compressor characteristic is shown in Fig. 4.2 where head is plotted against flowrate. As the flow is reduced the head and discharge pressure increases. The characteristic reaches the 'surge' point, or line, at a flow rate of about 70 or 80 percent of the best efficiency point with a pressure of about 120 percent of the rated pressure. At flows greater than the design flow the characteristic is limited by a rapid fall off of head. This is due to the high losses, particularly in the final stages in the compressor, caused by the high gas velocities and incidence angles at the entry into the impellers. This effect is accentuated as the Mach number at the inlet of the last impeller reaches one due to shock losses leading to choking or 'stonewalling' whence the head/flow curve becomes perpendicular. It is reasonable to limit the flow to 120 percent, or even, for machines with high pressure ratios, to 110 percent of the best efficiency flow with the head at 80 percent of the rated head. This point is known as the 'run out point'. At the surge point the head generated in the compressor is insufficient to deliver gas into the downstream system and a backward flow occurs. Operation is possible only if the process characteristic intersects the compressor characteristic to the right of the surge point. Any point to the left can only operate by bypassing some of the flow.

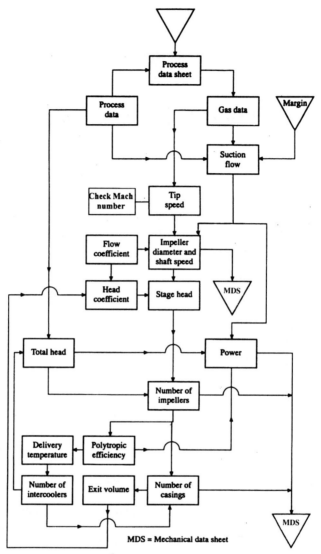

Fig 4.1   Centrifugal compressor selection flowsheet

4.3   Establish duties

(a) All process duties should be identified embracing the likely variations in inlet volume flow and head. This entails considerations of maxima and minima for mass flows, suction pressure, suction temperature, pressure ratio and the gas compositions to be compressed.

(b) Margins should be added to the higher duties to cover the following as applicable:

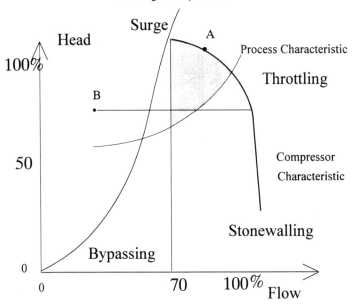

Fig 4.2 Characteristic of a centrifugal compressor

(i) The remaining uncertainties in process requirements and gas properties. These margins should be small if the likely duties have been fully developed already.

(ii) To enable the control system to restore equilibrium conditions after a disturbance has occurred.

(iii) To compensate for inaccuracies in performance measurement by the supplier. Compressor procurement standards permit no negative tolerance on head and flow. However, test codes generally permit an allowance for inaccuracies in the instrumentation.

(iv) To allow for machine deterioration in service. This may be wear of the internal leakage controls or process fouling or erosion.

(c) Whenever possible these margins should be assessed individually. However, if this is not possible, a single margin, covering them all should be applied. For applications where the pressure ratio can be defined fairly precisely, e.g. in some refrigeration services, a 10 percent margin should be applied to capacity. For applications where pressure ratio is heavily dependent on flow, e.g. a recycle duty, a 5 percent margin should be applied to capacity and the head. Higher margins than these can only be justified if the operating and capital cost increases are acceptable. In some cases extra margins are included in anticipation of future de-bottlenecking. This may be economic where variable speed drivers are employed.

(d) Pass-in sidestream pressures, within the compressor, should be specified at values below that anticipated in service to minimise the risk of the compressor generating high pressures backing out the sidestream flow. In the absence of better data, a value of 95 percent of the anticipated absolute pressure should be used. However, wherever possible, allow the compressor vendor some latitude in settling intermediate pressure levels to facilitate the selection of the most economic machine. Pass-in flows should be subject to the same margins as the first stage inlet flow discussed in 4.3(c) above. Particular care should be taken with sidestreams for compressors operating at different delivery pressures, e.g. refrigeration units where the final pressure depends on the ambient temperature.

(e) Conversely, pass-out sidestreams should be specified at values above that anticipated, say 105 percent of the absolute pressure.

(f) Intermediate pressures for intercoolers or multicasing compressors should not be specified at fixed values unless essential to the process. The compressor vendor should be permitted some freedom to optimise compressor selection. However, preliminary values will need to be assumed for initial assessment.

(g) Duties that have to be met only occasionally or for short periods of time must also be identified, e.g. commissioning, plant preheating, precooling or catalyst treatment. A common requirement is for precommissioning test runs on site, or off site such as a module yard, using a substitute gas such as air, nitrogen, CO_2, CO_2/He mixes, fuel gas etc. A suitable gas may need to be identified which does not lead to excessive power or discharge temperature and is readily and cheaply available. An alternative is to carry out precommissioning with the compressor under vacuum.

Occasional duties such as these should not necessarily be specified in full. The supplier should be permitted freedom to optimise the selection for the continuous process duties. The data sheet should show which parameters are mandatory, and which are desirable. For the latter, limiting values should be given if necessary.

(h) The data sheet should include all of the identifiable process duties (see 4.3(a) above). One of these should be designated 'normal operating point' at which usual operation is expected and optimum efficiency is desired. The vendor should guarantee flow, head, speed and power at this point.

One duty should be designated 'rated point' which defines the extremes necessary to meet all specified operating points. It will dictate the size and 100 percent speed of the compressor. This may be the normal operating point plus margins or any other operating point.

Additional duties, such as those mentioned in 4.3(g) above should also be listed.

Once all the duties, including any margins, are fixed it is helpful to plot them on an idealised characteristic of a shape described in Fig. 4.2. Assume the rated point to be at 100 percent flow and head, or slightly (2 percent) below. Check that all required duties are within the envelope of the characteristic and can be met.

Compressor control allows operation at points below the characteristic curve and to the right of the surge curve. This is done by either throttling the flow or modifying the characteristic of the compressor with guide vanes or speed control. By-passing is needed when operating to the left of the surge line.

4.4 Compressor configuration

A procedure for estimating the size, speed, number of casings, intercooling requirements and performance of multi-stage/casing centrifugal machines is outlined in the flowsheet of Fig. 4.1.

The 'rated point' is used as a basis to define the salient parameters of the compressor and identify any critical features. The method outlined can not be used to check part load behaviour, etc.

The procedure should first be applied to the total machine to obtain an overview. However, if more accurate data are required it can be applied separately, and slightly modified, to each section of the machine between intercoolers, and to each casing. The position of intercoolers and the number of casings may be assumed initially and analyses of re-runs will show if these assumptions are substantiated or not. If not they may be determined iteratively using the method described.

The procedure uses correlations between the inlet volumetric flow and compressor size (impeller diameter), compressor speed and compressor mean efficiency. The efficiency is then used to determine the compressor drive power. This enables the user to quickly determine the essential principal parameters, with an accuracy sufficient for design. It will also identify any compressor or application features requiring more detailed study.

4.5 Essential data

This should be for the rated duty.

Calculate the basic parameters from the data given (Chapter 2):

Suction flow	Q	m^3/s
Isentropic index	κ	
Compressibility coefficient at suction and discharge	z_s and z_d	
Molecular weight	M_w	
Speed of sound	u_a	m/s

4.6 Tip speed

Maximum tip speed is governed by the following:

(a) Strength limitation

The majority of process compressor impellers are manufactured from low alloy steels and tip speeds for closed impellers will be limited to approximately

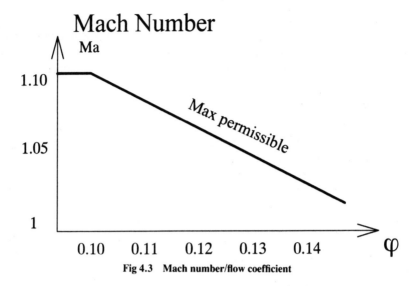

Fig 4.3   Mach number/flow coefficient

310 m/s. For open impellers, particularly overhung ones, tip speeds up to 400 m/s are acceptable. However, compressors handling corrosive gases, such as wet H_2S, require low strength materials to NACE (**15**). Assume a limit of 250 m/s.

(b) Mach number

Maximum tip speed is also limited by sonic considerations, therefore the Mach number inside an impeller, must be limited. The critical Mach number occurs at the eye of the impeller, and must be kept below about 0.8 to avoid choking at the inlet. (Inlet Mach numbers of up to 0.85 have been used.) However, it is extremely complicated to calculate this Mach number. For this reason the impeller tip Mach number, which is easy to calculate from a limited knowledge of the compressor, is used. The impeller tip Mach number is defined as

$$M_a = \text{Impeller tip speed/speed of sound at inlet conditions} \qquad (4.1)$$

The speed of sound of a gas can be calculated, see Appendix II (II.2).

The Mach number at the inlet is influenced not only by the tip Mach number but also by the flow coefficient and the hub ratio (hub/tip dia) of the impeller. A relationship has been developed between the impeller tip Mach number and that at the eye. Figure 4.3 gives the limiting tip Mach number for impellers with a hub ratio of 0.33. At higher hub ratios (0.4) the tip Mach number must be reduced by about 5 percent, at lower values (0.25) it can be increased by about 5 percent. The tip speed for closed wheels should not exceed 110 percent of the Mach number for flow coefficients up to 0.1 and then reduce to 100 percent as the flow coefficient increases to 0.14 (see Fig. 4.3).

For minimum capital cost select the maximum permissible tip speed. However, note that this leads to a narrow operation range. A wider range and a higher efficiency is obtained if the tip speed is slightly reduced (5–10 percent) though the number of impellers needed to achieve a given head is increased. Select a tip speed u.

4.7 Flow coefficient

It is economically desirable to use the smallest possible machine. The higher the flow coefficient the larger the suction rate for a given impeller diameter. For a multi-stage machine the first stage impeller should be chosen to have the maximum flow coefficient. The flow coefficient of the following impellers will decrease as their inlet volume decreases, if the impeller diameter is kept to the maximum and the shaft speed is constant. In multi-casing trains the first casing, because it has the highest suction volume, will dictate the train speed. This will lead to increasingly 'sub optimum' designs as the suction volume of the later stages decreases, which may be acceptable. However, for large pressure ratios (>10) it may be advantageous to use a higher shaft speed at higher pressures where the volumetric flow is smaller. In the limit compressors where one or a pair of impellers are driven by a pinion from a central bull wheel, allow a different speed to be used on every shaft, or even impeller. Note that this design requires a seal to atmosphere from each impeller. Overhung impellers are therefore used mainly on air and sometimes on compressors for inert gases.

The flow coefficients φ for closed impellers are in the range between 0.005 and 0.125, though impellers with larger flow coefficients have been used. The lower limit is dictated by the width of the impeller passage. The upper limit is dictated by the impeller shape or its Mach number. Manufacturers generally use standard designs arranged in a series of shapes and sizes. Each family covers a range of flow coefficients. Impellers with low flow and speed coefficients are one dimensional, impellers with high coefficient, particularly open impellers, are often three-dimensional. For a first attempt assume φ to be 0.06. For open impellers use 0.1.

4.8 Diameter and shaft speed

For a preliminary estimate calculate the impeller diameter of the first stage from

$$D = 2\sqrt{\{Q/(u\pi\varphi)\}} \text{ in metres} \tag{4.2}$$

and the shaft speed N from

$$N = u/(D\pi) \tag{4.3}$$

Note that manufacturers of centrifugal compressors do not produce machines with a continuous range of impeller diameters but rather have a series of sizes,

Table 4.1 **Pressure coefficients and polytropic efficiencies versus number of impellers for large volume flows**

Impellers/casing	—	2	4	6	8	10
Pressure coefficient	ψ	1.04	1.02	1.0	0.98	0.92
Polytropic efficiency	η	0.81	0.80	0.78	0.75	0.72

usually in steps of 20 percent, covering impeller diameters from about 0.25 to 1.5 m.

Normally the shaft speed is constant for the whole set. When calculating individual casings, once the overall number of casings has been established, the diameter and tip speed, as well as the inlet volume into the HP casing, will be known and the flow coefficient for the first impeller of the HP casing/s should be estimated from the formula given below

$$\varphi = 4Q/u\pi D^2 \qquad (4.4)$$

Where Q is the actual inlet volume rate into the casing considered. If the flow coefficient becomes too small it will be necessary to reduce the diameter of such wheels.

4.9 Polytropic index

In Appendix II the way to calculate the isentropic index κ is shown. The polytropic efficiency η is defined as:

$$\eta_p = (\kappa - 1)n/[(n - 1)\kappa] \qquad (4.5)$$

From that the polytropic index can be shown to be:

$$n = \eta_p \kappa/[1 - \kappa(1 - \eta)] \qquad (4.6)$$

Table 4.1 gives the polytropic efficiency against the number of impellers in a stage. It applies to large impellers ($Q > 0.5$ m³/s). At smaller flows the efficiency is reduced. See Fig. 4.4 where the maximum efficiency obtainable is plotted against suction flow.

4.10 Total head

(a) The head is calculated by assuming mean gas properties between suction and discharge values. Gas properties can be estimated using the method based on the ratio of critical pressure and temperature outlined in Appendix II. Note that this method is not accurate at high pressures and conditions close to saturation. Under such conditions one of the more complex methods should be used.

(b) Equation (A2.9) gives the formula for calculating polytropic head. Use

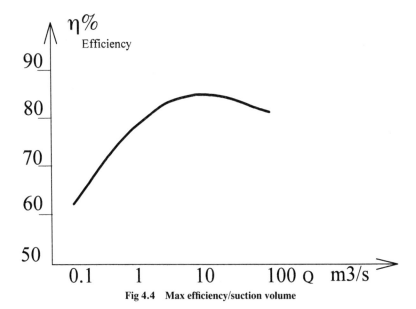

Fig 4.4   Max efficiency/suction volume

the polytropic index given in paragraph 4.6 above and use the average compressibility coefficient between suction and delivery

$$z_{av} = (z_s + z_d)/2 \qquad (4.7)$$

Calculate the total polytropic head and the delivery temperature assuming no intercooling. If this temperature is too high intercooling must be provided and the required total head must be increased by 1 and 3 percent for each water and air cooler respectively.

4.11   Pressure coefficient

Impellers with backward sloping blades have pressure coefficients in the order of one. This coefficient is a function of the impeller vane tip angle, the outlet velocity and the stage efficiency. It reduces at low flows.

To avoid very narrow impeller passages the vane tip angles for low flow impellers are made more acute so reducing the impeller radial flow velocity and allowing the impeller passage to be made wider. However, this reduces the pressure coefficient which is further reduced by the low efficiencies of such impellers.

Further the acute vane tip angle results in the actual exit vector to be nearly tangential which implies that operation is close to stalling and will rapidly lead to stalling if the flow is reduced. This is particularly critical for compressors operating with high pressure ratios: the flow in the last stage is proportionally much more reduced than that in the first impeller.

4.12   Stage head

The head developed by an impeller depends on its pressure coefficient. Take the pressure coefficient from Table 4.1. This gives the effect of impeller numbers on the pressure coefficient and on the maximum polytropic efficiency for closed impellers with backward sloping blades. Both efficiency and pressure coefficient are higher for open three dimensional impellers, particularly if they are overhung.

The stage head H is

$$H = \psi u^2/2 \tag{4.8}$$

Divide the total head required by the estimated head of a single impeller to obtain the number of impellers needed.

4.13   Number of impellers/casings

The number of impellers that can be accommodated in one casing is dictated by roto-dynamic considerations. These become more critical as bearing spans, rotor speeds and gas densities increase for a given compressor casing size.

Bearing spans increase with the number of impellers and their flow coefficients. Impellers with large flow coefficients have a greater axial length than those with small flow coefficients, any side branches (pass out or pass in) will also increase the bearing span and so reduce the number of impellers that can be accommodated in a casing. On multi-casing machines check if side branches into a casing can be avoided by arranging the machine such that side branches enter between casings.

The tip Mach number too has an influence on the number of impellers that can be installed in one casing. Figure 4.5 gives a guide to the permissible number of impellers based on Mach number, flow coefficient and tip speed. Table 4.2 is a guide based on practical experience.

The relationship between these parameters is complex and if the procedure outlined in the guide indicates that the maximum permissible number of impellers is needed for an application it is best to check with suppliers before finalising a specification.

4.14   Discharge temperature

The discharge temperature can be calculated if the pressure rise and the polytropic efficiency are known

$$T_d = T_s \Pi^{(n-1)/n} \tag{4.9}$$

The higher the gas temperature the higher the compression power. In the past 'isothermal compressors' with many cooling surfaces and intercoolers were

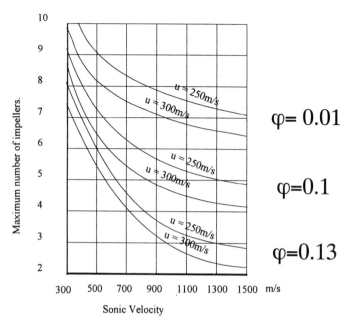

$$\varphi= 0.01$$

$$\varphi=0.1$$

$$\varphi=0.13$$

Sonic Velocity

Fig 4.5   Maximum number of impellers/casing

Table 4.2   Maximum number of impellers in a casing

Suction Rate m^3/s	Uncooled: no sidestream			Intercooled or one sidestream			Two sidestreams		
	<0.4	<4	>4	<0.4	<4	>4	<0.4	<4	>4
Pressure bar									
<70	10	8	7	9	7	6	8	6	5
70<120	9	7	6	8	6	5	7	5	4
120<200	8	6	5	7	5	4	6	4	3
200<350	7	5	4	6	4	3	5	3	2
300<500	6	4	3	5	3	2	4	2	—

popular. Due to their high capital cost they are rarely used now but by keeping the gas temperature low significant power savings can be made.

Mechanical considerations demand that the discharge temperature is limited to about 250 C. Special machines can be designed to have higher discharge temperatures, but the commonly used sealing elements made from elastomers, 'O' rings for the casing and in the oil system, usually limit the top

Table 4.3 Limiting temperatures for
some gases
These temperatures are based on a variety
of reasons, decomposition or reactions of
the gas handled being the most common
ones. Advice should be taken from a
chemist familiar with the gases to be
handled

	°C
Acetylene	60
Chlorine	100
Hydrocarbons	120
Ammonia	160
Carbon monoxide	180

temperatures to 170 C at rated conditions. Further, if the machine operates closer to surge a higher discharge temperature occurs. Cooled buffer gases are sometimes used to keep the temperature near sealing elements low. Some gases must be kept at lower temperatures. In those cases the permissible gas temperature limits the allowable discharge temperature (see Table 4.3).

4.15   Intercooling

If the temperature reached during compression exceeds the allowable temperature one or more intercoolers are needed. The temperature to which the gas is cooled depends on the cooling medium available (water, air or refrigeration). Special care must be taken when cooling air or other gases where water or other components can reach saturation and condense during cooling. If this happens the condensed liquid must be separated. In particular air, even if filtered, usually contains some dust which acts as a condensation core during condensation. If such condensate enters another stage the water evaporates and the solids settle on the impeller. When cooling below the dew point it is advisable to cool so that a fair proportion of the condensables are condensed. If only a small fraction is condensed separation of liquid and gas is inefficient and carryover is bound to occur. In such cases it is best to limit cooling to a temperature about 5 C above saturation.

4.16   Discharge volume

The minimum discharge volume rate for centrifugal impellers is about $0.03 \, \text{m}^3/\text{s}$ ($100 \, \text{m}^3/\text{h}$), provided that the flow coefficient at the discharge is above 0.005. This may require a lowered tip speed, by either reducing the shaft speed or the impeller diameter. This in turn leads to an increased number of impellers and a low efficiency.

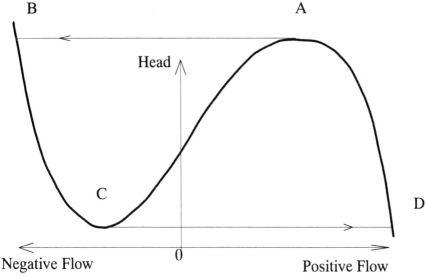

Fig 4.6   Characteristic curve

4.17   Power

For a preliminary estimate of power use the formula in equation (AII.12) with the head corrected for intercooling, and the mass flow and the efficiency from Table 4.1 and Fig. 4.4. For a more accurate estimate calculate the power for each stage. Add 5 percent for mechanical losses.

4.18   Compressor characteristic

In paragraph 4.2 the operating curve of centrifugal compressors is outlined. The total characteristic of a compressor is shown in Fig. 4.6. At the point of surge the flow switches from point 'A' to point 'B' and reverse flow occurs. This will lower the pressure at the delivery till point 'C' is reached. Here another unstable region is reached and the operation switches back to the normal characteristic at point 'D'. Note that this full cycle will only be traversed if the suction pressure is lower than that at point 'C' and 'D'. Further if a compressor is throttled directly at the delivery flange it is possible (on the test bed) to operate at points of the unstable curve between 'A' and 'C'.

Once surge is reached and if no action is taken the machine will operate in this cycle continuously. The frequency of the cycle depends on the volume at the delivery of the compressor. Surge can be very damaging to the compressor, cause excessive stresses in the impellers, damage to the thrust bearing and, if left unchecked for an extended period, lead to overheating of the compressor. Further, the unstable flow usually interacts with the process and therefore some action must be taken to prevent the compressor entering the region of surge.

4.19   Anti-surge

To prevent a compressor entering surge an anti-surge control is needed. This, in the simplest case is tripping the machine once surge is detected, a system favoured for some processes. More often a bypass, or blow off, valve is opened to maintain the flow above the surge flow. This action must be taken before surge is reached and many elaborate systems are in use. These control systems try to maintain flow at what is known as the 'anti-surge line'. Surge, as stated above, can result in damage to the compressor. Surge also results in an interruption of gas flow and so will affect the process. On air compressors a blow off valve is usually used but on gas compressors the flow is bypassed to suction after being cooled. Anti-surge systems have to be responsive to ensure that flow is maintained above the surge limit. Further they have to react extremely rapidly and their operation must not lead to hunting of the compressor. Computer simulations of the complete system, i.e. compressor plant and control instruments, are frequently required to design a satisfactory solution. Note that on recycle systems anti-surge is normally not required.

4.20   Rotating stall

This is closely related to surge. It occurs when the radial velocity of the gas entering the diffuser and/or the efficiency of the diffuser are too low then the flow as it passes through the diffuser can collapse into stall cells. Stall is more prevalent on low flow impellers where the radial velocities at the impeller exit are low. It is likely to occur on high pressure low flow applications. Stall tends to onset as the flow approaches surge. Using a diffuser passage narrower than the exit passage of the impeller reduces the flow at which stall tends to start.

Stall cells rotate at only 10 or 20 percent of shaft speed. The resulting vibration levels can be unacceptably high. Stall can also cause pressure pulsations in the delivery pipework.

4.21   Control

A compressor, without any internal control devices and operating at constant speed, will always operate at a point of the control line where this control line intersects the characteristic of the plant/process. The latter can be altered by introducing additional pressure drop with a suction or delivery damper, moving the operating point up the characteristic curve, so reducing the mass flow through the compressor albeit at the loss of efficiency (see Fig. 4.2 point 'A'). Note that throttling will allow operation only within the shaded area.

Opening a bypass will result in the flow to the process being reduced, again at the cost of efficiency (see Fig. 4.2 point 'B').

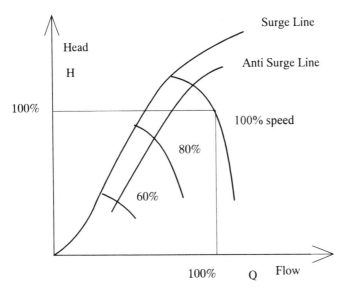

Fig 4.7 Speed variation

A more efficient method of control is to alter the characteristic curve of the compressor. This can be done by changing the speed—there is a separate curve for each speed (see Fig. 4.7). A similar effect can be achieved with guide vanes at the inlet of the first impeller. Note that sometimes more than one impeller is preceded with a guide vane particularly on integrally geared compressors.

A third method of altering the compressor characteristic is with variable diffuser vanes. However, this method is rarely used.

4.22 Operation

4.22.1 *Startup and shutdown requirements*

The compressor controls should be adequate for startup and shutdown. It is usual to keep the bypass valve fully open during startup and also open it when a machine is tripped. Tripping of a compressor should not drive it into surge. The reaction time of the anti-surge system should be rapid and the volume seen by the compressor at the delivery should be kept to a minimum, usually by mounting a fast acting non return valve near the delivery connection of the compressor, immediately downstream of the recycle of the line.

Reverse rotation of the compressor following a trip should be avoided especially when gas seals are used. A rule of thumb is that the energy of the gas entrained between compressor and check valve should be less than 50 percent of the kinetic energy stored in the rotating masses.

4.23 Roto-dynamics

The vibrational behaviour of rotors is the main limitation to the number of impellers that can be installed in one casing. There are two stimuli that lead to these vibrations: unbalance and fluid forces.

4.23.1 *Unbalance forces*

The rotor's response to out of balance forces must not lead to excessive forces at the bearings nor must the rotor be deflected to such an extent that internal clearances are lost and rubbing occurs. API 617 (**3**) specifies the maximum permitted out-of-balance forces and the acceptable responses to them. No lateral critical speed of the rotor supported on (flexible) bearings must fall within a defined margin of the operating speed range and further the response of the rotor at any speed below the operating speed should be such that no contact between rotor and stationary elements occurs. The calculated responses are often verified by shop tests.

4.23.2 *Fluid forces*

These forces originate in locations where the radial clearances between rotor and stationary parts are small, e.g. bearings and seals, as well as internal labyrinths when handling high density fluids. They can also emanate from shrunk on components when there is movement (microscopic) at their shrink interface. These forces must be controlled or alternatively sufficient damping must be introduced to prevent them generating 'sub synchronous whirl'. Such whirl occurs at a frequency equal to the first lateral critical speed of the rotor and can lead to rapid, within seconds, destruction of a machine.

Factors that increase the tendency to whirl are:

(a) The ratio of maximum continuous speed to 1st critical speed. This increases with increasing length of the rotor (number of stages). The larger the ratio the more likely whirl occurs.

(b) The density of the fluid handled. This effects the forces generated in balance pistons and other internal seals. Features that inhibit tangential swirl help to reduce cross coupling forces. Note that such features will normally only be needed at high pressures (>300 bar).

(c) Liquid buffered shaft seals can generate forces leading to whirl, though well designed seals have a beneficial damping effect. Dry gas seals work with an axial clearance and therefore do not generate, nor damp, such forces.

(d) To eliminate exciting forces from bearings tilting pad journal bearings are nowadays in common use. These bearings inhibit cross coupling forces.

(e) To avoid shrink fits fretting it is best to relieve such fits so that only a short axial length is subject to the shrink fit.

Various methods are used to estimate the magnitude and effect of cross coupling forces. However, these are all based on empirical data and comparison with successful designs and testing at full pressure is usually required by the operator to verify the response of ultra high pressure compressors (>180 bar).

4.24 Shafts

Forged low alloy steel shafts are standard for centrifugal compressors. Only machines handling highly aggressive gases have corrosion-resistant shafts. Shaft sleeves are frequently fitted so that sealing elements do not operate directly against the shaft.

To ensure trouble-free operation shafts must be carefully stress relieved to remove any internal stresses which could lead to deformation in use. Further, when shrinking on impellers, great care must be taken that when cooling down the impellers do not pull a bend into the shaft.

4.25 Impellers

Duties requiring one wheel are usually coped with by the use of a single overhung impeller with a step up gear part of the compressor. Overhung impellers are highly efficient: there is no need to accommodate a shaft in the eye of the impeller. It is for this reason that the overhung design is making inroads into air duties previously considered to be the area of single shaft compressors. Recently machines with up to six overhung impellers have become available. These machines have a slow running bull gear (1500 rpm) and two or three pinions each carrying an impeller on each end. One further advantage of this layout is that the pinion shafts can operate at different speeds so allowing each stage in a multi-stage machine to be run at its optimum speed. A disadvantage is that each impeller needs to be sealed towards atmosphere, so only applications where labyrinth seals are acceptable can be met.

Impellers with a high flow coefficient are frequently open. Others are usually of the closed design. In the past all impellers were made from forgings with the vanes riveted in. Modern design favours cast, welded, and also brazed impellers. Impellers with very low flow coefficients may have internal channels spark eroded, slot welded blade covers, or may be brazed.

Impellers are shrunk onto the shaft, either hot or hydraulically in the case of single shaft machines. Internally-geared machines may have the impeller shrunk onto the pinion shaft, though more commonly they are fixed with a Hirth or similar coupling to facilitate assembly. Some compressors use polygon-fitted impellers.

4.26 Seals

To prevent leakage from high pressure to low pressure regions seals are needed. Internally shafts and impeller eyes are always sealed with labyrinth seals. These are also used to prevent gas escape on machines handling air where leakage into the atmosphere can be tolerated. Compressors handling toxic or flammable gases need positive seals. At low pressures oil flooded mechanical seals can be used, at higher pressures floating ring seals or dry gas seals are required. See Chapter 14 for details.

4.27 Bearings

To ensure long-term reliability vibration-free running is required. Journal bearings have a significant influence on the roto-dynamic behaviour of shafts and must be carefully chosen. Tilting pad journal bearings are nearly always used on today's machines. Magnetic levitation bearings in combination with dry gas seals and high speed motors, have been used on oil-free compressors. Their applications are increasing, particularly for machines installed in remote locations.

4.28 Axial unbalance and bearings

In operation there is a pressure rise across each impeller. This pressure rise leads to an axial force, and this force must be resisted by an axial (thrust) bearing. For low pressure machines this presents no problem, but for high pressure compressors the thrust generated is much greater than can be handled in a thrust bearing. Thrust forces must, therefore, be balanced, either by having a double entry or back to back impellers, so that the gas flow inside the machines is partly in one direction and partly in the other, or with the use of a balance piston (see Fig. 4.8).

Unbalance Force = A $(P_d - P_s)$ Balance Force = A$_{BP}$ $(P_s - P_d)$

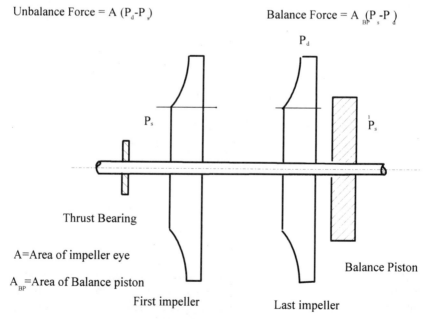

Thrust Bearing

A = Area of impeller eye

A$_{BP}$ = Area of Balance piston

First impeller   Last impeller

Fig 4.8   Axial forces (simplified)

Thrust bearings can carry a load of up to 20 bar and can operate at a maximum rubbing speed (tip speed) of about 110 m/s. On high pressure compressors this means that only a small fraction of the axial force can be carried by the thrust bearing, the rest must be 'balanced'. This often leads to problems as the remaining force is the difference between two forces which are both a magnitude larger. Double acting tilting thrust bearings are the norm. Frequently these incorporate a device to equalise the load on the pads.

Integrally geared compressors have no thrust bearings on pinion shafts. The thrust is partly balanced, in the case of two impellers mounted on one pinion, and also balanced against the helical gear. Axial location is by a thrust face on the pinion facing the bull gear. The latter has a double acting thrust bearing.

4.29 Casings

In paragraph 4.13 the type and number of impellers needed is calculated. What sort of casing is required?

Horizontally-split casings are standard on low pressure machines except if a barrel casing is demanded by API 617 (**3**) or other relevant code to fulfil containment requirements for light gases. Cast iron is the usual material, cast steel being required when hazardous gases are being handled.

Vertically split or barrel casings, in cast steel or forged, are generally accepted for high pressure machines (>50 bar). The heavy bolting required by horizontally split casings is avoided. Parts can be circular and sealed with 'O' rings, and pressure forces are more and more held with shear rings rather than with bolted on end covers. The diffusers and return passages are made from cast iron, nodular iron or steel in both horizontal and vertical split machines, depending on the pressure differentials and diameters. These parts, except for overhung machines, are always made in semi-circles. For barrel compressors they are assembled around the impellers and the complete assembly, the 'bundle' is then dropped or slid into the casing. To prevent leakage through stationary parts these are sealed with 'O' rings or similar materials.

4.30 Vessels

Centrifugal machines, in contrast to reciprocating machines, do not require special designs for vessels. However, care must be taken that piping between coolers, catchpots and compressors do not cause excessive loads which lead to distortion and misalignment of compressors. See paragraph 4.34, Layout.

4.31 Instrumentation

Instrumentation for centrifugal compressors has two separate facets: process and mechanical. The former needs measurement of inlet and delivery conditions, i.e. temperature and pressure for every stage. With these readings the polytropic efficiency can be calculated separately for each stage (see equation

(AII.13)). A reproducible flow indication further allows the head developed in each stage to be plotted over a period to check for any deterioration, which is a great help in diagnosing the condition of a stage. However, in the process industry many compressors are operated at near constant flow, making flow measurements for condition monitoring desirable but not essential.

Instrumentation to prevent surge is discussed in paragraph 4.18.

Vibration measurement is nowadays universally installed. For integrally geared machines external accelerometers are the best indication of changes of the mechanical condition of the unit. For single-shaft machines, particularly barrel casings, proximeters measuring the shaft displacement relative to the stationary bearings is the most meaningful indication. Readings are indicated and used to operate alarms, sometimes trip, when excessive vibrations are reached. Further each axial location bearing should be 'supervised' by an axial sensor arranged to trip the driver once the limit of movement has been reached. Bearing temperature measurement is usually provided through elements embedded or inserted in the bearing white metal lining. Readings are indicated and are used to operate alarms and sometimes trip when excessive temperatures occur.

4.32   Safety and availability

Well-designed centrifugal compressors have a high availability and are not prone to leakage. The energy stored in rotors is relatively low, so that even if a rotor or part of it would split the fragments would not penetrate the casing. This does not apply to couplings. If these fail they invariably penetrate the coupling guard. Special care is therefore needed in selecting, inspecting and operating couplings for high-speed machines. About 30 percent of all outages of centrifugal compressors are caused by bearing problems. Because of their high availability and reliability centrifugal compressors are favourites for single stream plants.

4.33   Couplings

Integrally geared machines have only one low speed coupling, situated between the driver and bull gear. A coupling containing elastonomers is frequently used.

Single-shaft machines require high-speed couplings. In the past oil-lubricated gear couplings were favoured. Today flexible dry element couplings are replacing gear couplings. These can be designed such that in case of failure of the membranes or diaphragm they become a crude gear-coupling allowing safe shutdown, albeit with increased vibrations.

Some manufacturers prefer solid couplings or torsion rods between driver and compressor. This has the advantage for geared drives of eliminating a high-speed thrust bearing, the thrust of the high-speed shaft being transmitted through collars on the pinion to the low-speed shaft thrust bearing.

4.34 Layout

4.34.1 *Branch orientation*

Most centrifugal compressors have a horizontal shaft, though some smaller single stage units are being built with a vertical shaft with the motor above the compressor. The pipe loads that can be imposed on a casing are limited (see NEMA SM23 (**16**)) and rigid anchor points are rarely available above machine level. This means that even if the machine branches point upwards the pipes will have to be brought down to be attached to rigid anchor points, though the weight of pipes can usually be supported above a machine.

Major maintenance on horizontally split casings is very much simpler with downward pointing connections: there is no need to disturb the pipe work when opening the casing to attend to the rotor. Compressors with vertically split casings, with a free end, can be stripped and maintained with the casing in position—only the 'bundle'—the inner assembly of rotor and diffusers—being 'pulled'. Vertically split casings in the middle of a train must, however, be removed before they can be dismantled.

Single and double stage gas line boosters are usually erected at a low level with inlet and outlet opposite each other in the horizontal plane or with axial suction and a side discharge.

Inter and after coolers, as well as vacuum condensers for turbine drivers are conveniently located below the operating floor so calling for downward pointing branches. For these reasons downward-pointing branches are preferred. However, upward-facing nozzles are used in many offshore compressor installations.

4.34.2 *Pipe forces*

The forces and moments that can be allowed to act on machine casings are limited for a number of reasons: they can distort the casing, causing internal misalignment and loss of clearance, lift the casing off its supports, resulting in external misalignment and in the limit deform or break holding down bolts.

NEMA-SM23 (**16**), a standard developed for steam turbines, is the accepted criteria for calculating permissible pipe loadings. The more robust machines are usually designed to be able to support forces and moments 1.85 times of those calculated according to NEMA SM23. However, there are some machines, particularly machines designed to withstand rapid changes in temperature, where the pipe loads must be limited to much lower values, such as half the allowable values given in NEMA. To ensure trouble free operation it is essential that acceptable pipe loads are agreed with the machine vendor.

It is simple to calculate and support the weight and the dead loads exerted by the pipe work. Calculating forces due to thermal expansion is much more involved and is best done with the help of a computer. Particular attention must be paid to ensure that nominal rigid points are truly rigid and that support points with one or two degrees of freedom are free to slide in the specified directions.

Occasionally loads caused by temperature changes are reduced or halved by designing the pipe work such that zero load occurs at half the operating temperature. This is done by including 'cold pull' or 'set' into the design. Though this helps to limit pipe loads, it is a practice to be avoided. Maintenance personnel are usually unaware of the purpose of the pipes not meeting and modify them so taking the cold pull out. Further it is very difficult to check on site the amount of cold pull applied.

Pipe load calculations must be carried out during the design. Erection personnel can only check that the flanges meet correctly—they are not qualified to estimate pipe loads.

4.35 Drivers

Ideally the compressor and driver speeds should match without the need for an interposing gear. However, this is not always possible.

(a) Electric motors

Motor drives require, with the exception of very large units with high inlet flows, that a step up gear is used. Variable speed motors are limited in power (see Chapter 12).

(b) Steam turbine

Standard designs are available for mechanical drives. The speed of back pressure designs can usually be matched to the compressor speed. Condensing turbines may need to have multiple exhausts to match compressor speed without exceeding blade stresses at the exhaust end.

(c) Gas turbine

The speeds of gas turbines are standard for a given frame. Examine if the output speed of the gas turbine can be used to design an efficient compressor. If not an intermediate gear will be needed.

4.36 Example

160 000 Nm³/h of synthesis gas has to be raised from 26 bar to 80 bar. The gas has a molar mass of 9 and is at 20 C. Its ratio of specific speeds is 1.4. Figure 2.2 shows that a centrifugal machine is suitable to fulfil this duty.

> Suction flow: 1.84 m³/s allow a 5 percent margin. Total suction flow Q = 1.93 m³/s
> With such a light gas a high tip speed can be chosen 310 m/s, giving a Tip Mach No. of 0.51.

From equation (AII.9) the total head H is

$H = (8.314/9) \times 293 \times 1.4/(1 - 1.4) \times (\Pi^{(1.4-1)/1.4} - 1) = 358$ kJ/kg

Choose a pressure coefficient of 0.8. The head developed by one impeller becomes

$H_{\text{stage}} = 0.8(310)^2/2 = 38$ kJ/kg

A 10 stage machine is needed. With this number of impellers a 2 casing machine is called for.

From equation (4.2) the diameter D is

$D = 2 \sqrt{\{1.93/(310 \times \pi \times 0.06)\}} = 0.36$ m

This gives a flow coefficient φ at the inlet of

$\varphi = 1.93/(310 \times 0.36^2 \times \pi/4)$

$= 0.061$

The flow coefficient at the last impeller, neglecting any temperature correction will be

$= 0.061 \times 26/80 = 0.02$

Both values are acceptable.

From equation (4.3) $N = 310/(\pi 0.315) = 271$ Hz (16 300 rpm)

Polytropic index n.
From Fig. 4.4 find the efficiency of the 1st stage = 0.78 the higher stages have a smaller volume flow so assume the overall polytropic efficiency to be 70 percent.
Using equation (4.6)

$n = 0.7 \times 1.4/[1 - 1.4(1 - 0.7)] = 1.7$

The delivery temperature T_d (4.9) $= 293 \times (80/26)^{0.7/1.7} = 465$ K (192 C)
This is a higher temperature than tolerable with most gases. An intercooler between the two casings will, therefore, be required.
Total power mass flow: 160 000/(3600 22.4) 9 kg/s = 18 kg/sec
Polytropic head

$H = 8.314/9 \times 1.7/7 \times 293 \times \{(80/26)^{0.7/1.7} - 1\} = 387$

From equation (A2.14) the power P is

$P = 18 \times 387/0.7 = 10$ MW

An intercooler will lower the required power by about 5 percent.
A two casing machine operating at about 16 000 rpm would be suitable.

Chapter 5

Axial Compressors

5.1 Introduction

Chapter 5 covers axial compressors. It is assumed that you have already been through the process of preliminary selection (see Chapter 2) and an axial compressor is the appropriate choice for the duty required.

The high efficiency obtainable with axial compressors makes them a favourite when large rates are required (see paragraph 2.4.1(a) and Table 2.1). Further they are considerably smaller than centrifugal compressors of similar capacity. They appear to be of simple design, but this is not so. While it is simple to calculate overall dimensions, the detail design of blades, their shape, size, spacing and reaction is the subject of many books and articles based on theory and practical experience. 'Turboblowers' by Stepanoff (**17**) gives a good introduction. Axial compressors are more complicated than centrifugal machines and are not able to withstand the misuse a centrifugal machine can tolerate. In particular they are much more likely to be damaged by surge and they are much more affected by dirt and liquid carryover. They are used mainly as air and LNG gas compressors.

5.2 Characteristic

The characteristic of axial compressors is limited in a similar way to that of centrifugal compressors with surge on one side and stonewalling at maximum flow. However, while in the latter the operating flow range is about +15 percent the operating range of axial machines is much narrower (see Figs 5.2 and 5.3).

To be able to match a given duty accurately the guidevanes on fixed speed compressors are frequently arranged so that their angle can be adjusted automatically in service or once the compressor has been opened up. Great care must be taken to ensure that all guideblades on one stage are set to the same angle.

5.3 Establish duty

Axial compressors are usually not required to fulfil a number of different duties. It is therefore much simpler to establish the duty than it is for centrifugal compressors (see paragraph 4.3).

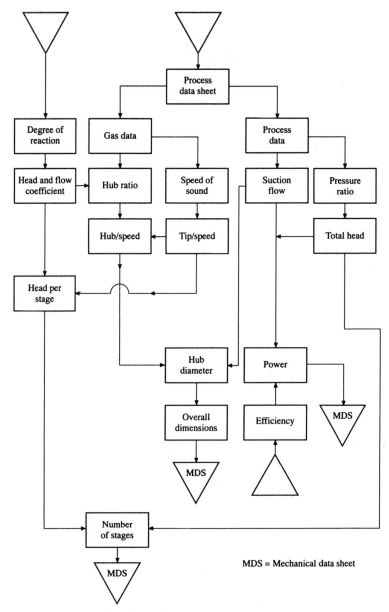

Fig 5.1   Axial compressor selection chart

5.4   Design of axial stage

An axial compressor consists of a number of rotating and stationary cascades. The rotor blades are carried by the, usually, cylindrical rotor. The stator is tapered towards the exit as the volume flow decreases with increasing pressure.

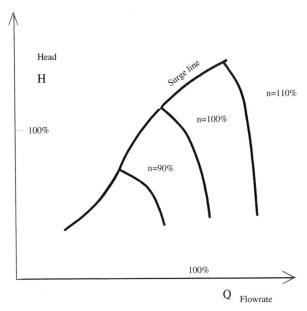

Fig 5.2 Characteristic of an axial compressor with speed control

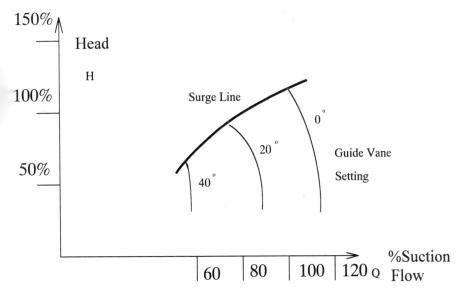

Fig 5.3 Characteristic of axial compressor with variable guidevanes

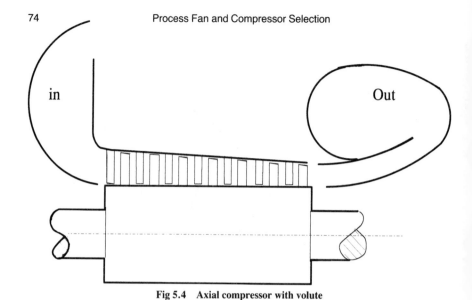

Fig 5.4   Axial compressor with volute

The rotor blades have a reaction between 50 percent and 100 percent, depending on the application. Machines with 50 percent reaction have a larger flow coefficient and operate at higher speed, hence have smaller overall dimensions and are lighter: both advantages for aviation gas turbine compressors. Machines with 100 percent reaction have lower blade stresses and are less sensitive to fouling and as there is little change of pressure over the stator blades large tip clearances of the stator blades do not significantly drop the efficiency. Further, due to their lower velocities, the design of energy recovery after the final rotor stage is simpler.

5.5   Essential data

This is the same as that required for centrifugal machines (see paragraph 4.5).

5.6   Arrangement

For compression ratios up to five a single section, or seven if terminated with a centrifugal stage, is possible (see Figs 5.4 and 5.5). For higher compression ratios two sections, frequently terminated with radial stages and with an intercooler are selected (see Fig. 5.6). Termination of an axial compressor with a radial stage dispenses with the bulky diffuser otherwise needed, so reducing the axial length of the rotor and consequently of the machine.

5.7   Sizing

The gas has to pass through the annulus between rotor and stator. In this guide all velocities are referred to the hub diameter. With the common cylindrical

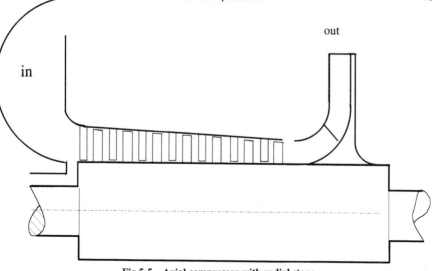

Fig 5.5   Axial compressor with radial stage

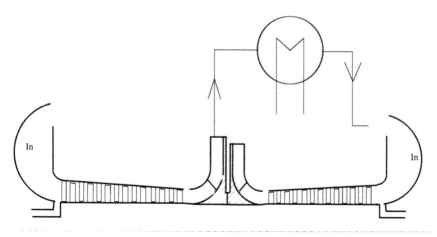

Fig 5.6   Two stage axial compressor

rotor this results in only one diameter 'd_{hub}' and one speed 'u_{hub}' to be used in all calculations. Note that using 'u_{hub}' and 'd_{hub}' is not universal. Other publications use 'd_{mean}', '$d_{average}$' and 'd_{outer}'. Both flow and pressure coefficient alter if they are based on a different diameter.

The maximum diameter of the rotor is restricted by two parameters: centrifugal blade stresses, limiting the hub velocity to 185 m/s, and the need to keep the tip velocity below the speed of sound. For air the maximum allowable tip velocity is about 300 m/sec, an M_{tip} of 0.9. The maximum permissible diameter ratio D_r for air is 0.6 so the hub Mach number $M_{hub} = 0.54$. For LNG D_r is about 0.68.

The flow coefficient φ, defined for axial compressors as

$$\varphi = c_{axial}/u = Q/Au \tag{5.1}$$

where c_{axial} the gas velocity in the direction of the shaft and A the area of the annulus at the inlet.
φ is in the order of 0.7 for machines with 50 percent reaction and drops to 0.5 for blading with 100 percent reaction.
With a hub diameter ratio D_r the hub diameter d becomes

$$d = \sqrt{\{4Q/\pi\varphi u(1/D_r^2 - 1)\}}\,\text{m} \tag{5.2}$$

Assume the flow coefficient φ to be 0.7.

The number of stages needed depends on the required head. The head of a single stage can be estimated from the hub velocity and the pressure coefficient ψ.

The pressure coefficient of axial compressors is between 0.5 and 0.6 and 0.4 and 0.45 for 50 percent and 100 percent reaction, respectively.

$$H_{stage} = \psi u^2/2 \tag{5.3}$$

The number of stages E required becomes

$$E = 2H_p/\psi u_{hub}^2 \tag{5.4}$$

The internal power is the product of mass flow m by the polytropic head divided by the polytropic efficiency (see equation AII.12)

$$P_{int} = mH_p/\eta_p \tag{5.5}$$

The polytropic efficiency can be has high as 90 percent and 87 percent for air compressors and LNG compressors, respectively. The maximum stage efficiency is higher but depends on the hub ratio, and as this decreases at the high pressure end the stage efficiency drops.

5.8 Control

Axial machines have a steep characteristic. For this reason rate control by throttling—always a method with poor efficiency—is not feasible. There are two methods of control: speed variations and adjustable guide vanes. The latter requires that the guidevanes on some or even all stages, depending on the degree of reaction, have to be adjustable from the outside, giving a 'variable geometry'.

5.9 Surging

Axial machines are especially sensitive to surging. Not only is the blade load during surging a multiple of the normal load but the hot gas flowing back to the inlet and then being recompressed leads to a rapid increase in temperature in the bladed section. A reliable bypass/blow off valve is essential. It is common on chemical plants to install surge detectors and, in case of surge, to trip the compressor after only one, sometimes two, surge pulses are detected.

5.10 Casing

Axial compressors are low pressure machines usually handling air. Casings of industrial designs are always horizontally split. Stator blades are usually mounted on an inner casing. Cast iron is the normal material for outer casings on air duties.

5.11 Overall dimensions

The most characteristic dimension of an axial compressor is its hub diameter. The outside diameter of the casing is about six times the diameter of the hub. The length of a single stage compressor is about six to eight times its hub diameter.

5.12 Drivers

Axial machines are large capacity machines and these by necessity have large power demands. Electric drives are used if the supply network is adequate, in particular it must allow safe starting. Gas and steam turbines are used if the electric supply is inadequate. The latter have the advantage that they allow speed control.

5.13 Example

42 m³/s (150 000 m³/h) are required at a pressure of 4 bar. Suction is atmospheric and the suction temperature is 25 C. Figure 2.2 shows that an axial compressor is the most suitable choice for this duty. The speed of sound of air at 25 C is 347 m/s. The maximum allowable tip velocity is 90 percent of the speed of sound, i.e. 310 m/s. The compressor will also have to operate during the winter when the speed of sound will be lower. A tip velocity of 300 m/s and 50 percent reaction is chosen, giving a flow and pressure coefficient of 0.7 and 0.55, respectively. The hub velocity, with a diameter ratio D_r of 0.6 becomes 180 m/s.

Using equation (5.2) calculate the hub diameter d

$$d = \sqrt{[4QD_r^2/\{\pi\varphi u(1 - D_r^2)\}]}$$

$$d = \sqrt{(4 \times 42 \times 0.36/\pi \times 0.7 \times 180 \times 0.64)} = 0.49 \text{ m}$$

The tip diameter of the first stage becomes 0.82 m.
From equation (5.3) the stage head H_{stage} is

$H_{stage} = 180^2 \times 0.55/2 = 8.9$ kJ/kg

The total polytropic head required, assuming an efficiency of 85 percent becomes:
From equation (4.6) the polytropic index n is

$n = 0.85 \times 1.41/(1 - 1.41 \times 0.15) = 1.52$

From equation (AII.9) the polytropic head H_p is

$H_p = 8.314 \times 1.52 \times 298/(29 \times 0.52) \times (4^{0.52/1.52} - 1) = 152$ kJ/kg

From equation (4.9) the number of stages required

$E = 152/8.9 = 17$

From equation (4.9) the delivery temperature

$T_d = 298 \times 4^{0.52/1.52} = 479$ K (206 C)

The air volume flow of 42 m³/s corresponds to a mass flow of 50 kg/s.
From equation (5.4) the power

$P = 50 \times 152/0.85 = 9$ MW

From equation (4.3) the shaft speed

$N = 180/\pi \times 0.49 = 117$ Hz (7015 rpm)

The footprint of the compressor will be approximately 3 × 4 m

Part Three

Positive Displacement Machines

Chapter 6

Reciprocating Compressors

6.1 Introduction

Chapter 6 covers reciprocating compressors. It is assumed that you have been through the process of preliminary selection (Chapter 2) and a reciprocating compressor is the appropriate choice for the duty required.

Many conditions affect the satisfactory performance and availability of a reciprocating machine. The design is not only based on mechanical considerations but also on the gas or gas mixture to be compressed, which not only affects the allowable pressure ratios, but may cause corrosion or erosion, reduce the properties of lubricants, cause fouling and blockages, produce condensables or promote explosive mixtures and may require special metallurgy.

The following pages will discuss in brief detail some of the above points. When handling aggressive, noxious or flammable gases the machines engineer should consult on the problems involved with a metallurgist, and process and safety engineers to ensure a safe machine with high availability and cost effectiveness.

6.2 General description

Reciprocating compressors cover a wide range of applications as shown in Chapter 2. In this book only large machines are considered. Small compressors, such as used, for example, for compressed air in garages, are not dealt with.

There is a great similarity between reciprocating compressors and internal combustion engines—both have a number of cylinders, varying from one to eight, arranged in line, V formation or horizontally opposed; each type of machine has a crankshaft, cylinders, pistons and connecting rods. Two major differences are that an engine needs to have its valves positively operated whereas a compressor valve is self-acting, i.e. it is a pure non return valve. The other difference is that in all but the smallest and refrigeration compressors with pressurised crankcases, the connecting rod is not directly acting on the piston but on the crosshead which in turn is connected to the piston with a piston rod (see Fig. 6.2). This feature has many advantages, among them are

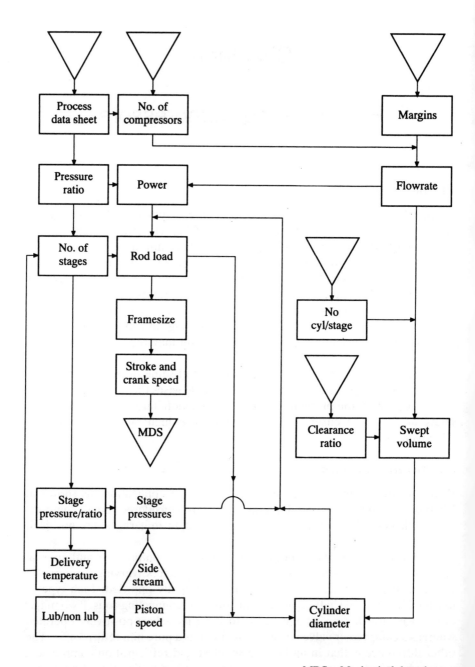

MDS = Mechanical data sheet

Fig 6.1   Reciprocating compressor selection chart

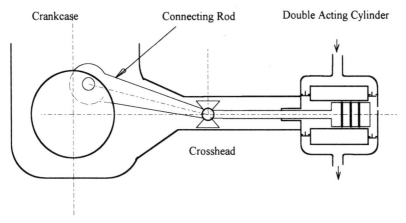

Fig 6.2 Reciprocating compressor

that the cylinder is isolated from the motion work and can be lubricated with a special lubricant or even be unlubricated. The crosshead design takes all sideways thrust away from the cylinder wall—the thrust is taken by the cross head, which is designed and lubricated to do just that. Further the use of a piston rod allows the cylinder to be double acting, i.e. gas can be compressed on the outward stroke as well as on the inward one.

6.3 Basic process data

In Appendix II it is shown how to calculate the actual volume given a mass flow and gas conditions. Use the most severe of the data sets provided on which to base the machine design. This is usually the set with the largest mass flow and the minimum suction pressure as well as the maximum inlet/interstage temperature.

6.4 Number of compressors

Before you can proceed with a compressor selection you must decide on the number of compressors to be used i.e. does the duty require one 100 percent machine, two, or even three 50 or 60 percent units? Note that, although the availability of reciprocating compressors is fairly high, 'non-availability' mainly due to damaged valves, occurs at random. To replace a damaged valve takes only a short time (between 1 hour and $\frac{1}{2}$ a day); however many applications require a constant supply of gas (e.g. instrument air) and even a brief interruption of supply can lead to an extended shutdown of the process the compressor serves.

Note that infrequently used spare machines may deteriorate and may need special measures to ensure maximum availability.

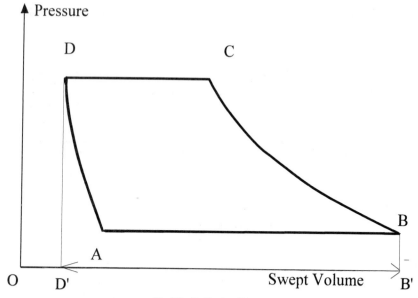

Fig 6.3   **Indicator diagram**

Guidance on the number of compressors required is given in Chapter 2, Table 2.2. Once the number has been fixed the selection can proceed.

6.5   Basic design

Process design is generally based on a mass rate, while compressor design is based on a volume flowrate at suction conditions, pressure ratio and delivery pressure. The parameters that define a reciprocating compressor are the number of stages and frame size. The latter is based on the allowable rod load, i.e. the maximum force that can be transmitted through the piston rod. This is a convenient quantity specifying the loads and stresses a compressor's parts are subject to.

The number of stages required is related to the permissible pressure ratio, which is not only limited by the rapid reduction in volumetric efficiency with increasing pressure ratio, but also by the maximum allowable temperature in any one stage. In small single stage machines pressure ratios up to four can be used but in larger and multi-stage compressors the maximum is usually less than three.

The action inside a cylinder of a reciprocating compressor is best shown with the help of the indicator diagram (see Fig. 6.3). Suction starts at point 'A' and at dead centre 'B' the piston reverses and suction stops. This is followed by the compression stroke. The volume of the gas is reduced and its pressure increases until the pressure inside the cylinder exceed that in the delivery line when the delivery valve opens (point 'C'). Gas is then expelled from the cylinder till the piston reaches dead centre and the delivery valve closes at point 'D'. The gas remaining in the cylinder (in the clearance volume) will then expand till its pressure is below that in the suction line when the suction valve will open at point 'A', so completing the cycle. The ratio of OD′/D′B is the clearance ratio 'ε'.

Power consumption is directly proportional to the volume of the gas, and as the volume of a gas decreases with decreasing temperature, the lower the temperature during compression, the less the power required for a given mass flow. Isothermal compression (compression at constant temperature) is the ideal case but can not be achieved in practice. However, by keeping temperatures as low as possible at all times the required compression power is kept to a minimum. Cooling cylinders is effective on small compressors but in large cylinders compression approaches isentropic conditions. In a multi-stage machine the gas is cooled after each stage so the ideal, i.e. isothermal compression is approached. This would suggest that the more stages the better, but as there are some pressure and mechanical losses associated with each stage, as well as increased capital cost, a stage compression ratio between two and three is found, in practice, to be the optimum.

6.5.1 Delivery temperature

The permissible delivery temperature is limited by the cylinder lubricant and by the gas handled. Table 6.1 gives some guidance.

The maximum permissible pressure ratio Π of a stage can be estimated once the delivery temperature is fixed, using the following formula:

$$\Pi = (T_d/T_s)^{n/(1-n)} \tag{6.1}$$

where n is the polytropic index of the gas handled.

Assume n to be equal to the ratio of specific heat κ for a first estimate. There is always some heat transfer between the cylinder and the gas. Heat transfer in a cylinder is a complicated process. During the suction stroke the gas is heated by mixing with the residual gas that had remained in the cylinder and by heat flow from the cylinder walls. During compression the heat flow reverses and as the gas gets hotter some heat flows into the cylinder. This flow of heat into the

Table 6.1 Permissible gas delivery temperatures in compressors
These temperatures are based on a variety of reasons: decomposition, or reaction of the gas handled being the most common ones

Gas	C	
Air	160	Lubricated (to avoid cylinder explosions and fires)
Hydrogen	200	
Nitrogen	190	
Chlorine	<100	Corrosion
Unlubricated	<200	Depends on ring material

cylinder continues till the flow again reverses at some time during the expansion stroke. For cooled cylinders the polytropic index can be estimated with the help of the following empirical formulae:

$$n = (1 + \alpha)\kappa/(1 + \kappa\alpha) \tag{6.2}$$

where α is the warm up factor

$$\alpha = 0.15/(DN^{1/2}) - 0.4\varepsilon \tag{6.3}$$

where ε is the clearance volume

Note that n is usually smaller than κ particularly for cylinders with a small bore. A simplistic view is that both the compression and expansion stroke are adiabatic. However, the heat transfers that do occur cause a continuous change in these processes and the above formula for polytropic compression is only an approximation based on the conditions in points 'A', 'B', 'C' and 'D' in Fig. 6.3.

The ratio of AB/D'B, i.e. the suction volume divided by the swept volume is known as the indicated volumetric efficiency η_v and can be calculated from

$$\eta_v = 1 - \varepsilon[\Pi^{1/n} - 1] \tag{6.4}$$

The actual volumetric efficiency η_{va} is lower between 2 percent for low compression ratios (about two) to 20 percent for large compression ratios. The actual value being influenced not only by the compression ratio but also by the design of the piston and the ratio of specific heat κ of the gas handled (18).

6.5.2 Number of stages

The volumetric efficiency, as already mentioned above, of a cylinder decreases and the delivery temperature increases with increasing compression ratio. Further the power consumption increases with increasing compression ratios. Figure 6.4 shows the relative power consumption against compression ratio with the number of stages as a parameter.

For these two reasons it is necessary to use multi-stage machines when a large compression ratio is required. A preliminary estimate of the number of stages E required can then be obtained from:

$$E = \log \Pi_0/\log \Pi \tag{6.5}$$

where

Π_0 The overall pressure ratio and

Π The stage pressure ratio

Round up to the next higher integer.

In most cases if more than one stage is required intercooling will be needed. The temperature exit the intercooler or coolers is often higher than the suction temperature of the first stage, particularly in summer. To avoid excessively high delivery temperatures in the higher stages it may be necessary to limit the

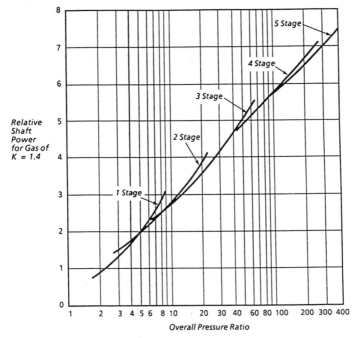

Fig 6.4 Effect of number of stages on absorbed power

permissible pressure ratios in these stages before proceeding with the configuration of the compressor. Check if the additional demands made when rate control is required are met. In general any control system operating on the first stage only will result in the compression ratio and the delivery temperature of the last stage increasing as the rate is decreased. It is therefore necessary to ensure that the last stage has only a small compression ratio at maximum throughput so that at lower rates the permissible delivery temperature is not exceeded in the last stage.

6.5.3 Power

The isothermal power can be estimated by the following formula:

$$P_{is} = 100Qp_s \lg p_d/p_s = 100Qp_s \lg \Pi \quad kW \tag{6.6}$$

At higher pressures the effect of compressibility cannot be neglected and the formula becomes

$$P_{is} = 100Qp_s \lg \Pi \sqrt{z_s z_d}$$

For a first estimate of the total power (P_{tot}) required at the coupling assume an efficiency of 65 percent.

$$P_{tot} = 154Qp_s \lg p_d/p_s = 154Qp_s \lg \Pi \quad kW \tag{6.7}$$

$$P_{tot} = 154Qp_s \lg \Pi \sqrt{z_s z_d} \quad kW \tag{6.7a}$$

6.6 Throughput control

Most reciprocating compressors run at constant speed and have roughly a constant inlet volume flow even if the overall pressure ratio varies.

Below, some of the ways used to control the throughput, are outlined:

6.6.1 *Inlet line throttle valve*

This method is wasteful of power but useful to control exhausters with sub atmospheric inlet. The control range is less than 2:1. It is also used to operate compressors where approximately constant mass flow is required even if large variations in suction conditions occur.

6.6.2 *Inlet line cut-off valve*

This is an 'On Off' control and so gives regulation over the full flow range. The compressor is not stopped, but is unloaded, operating at 100 percent or zero flow. To keep the pressure in the delivery system constant and smooth out flow variations a large receiver is required. More than one compressor can be used to supply the receiver though this increases the pressure variations in the system as each compressor has an individual pressure sensor to start and stop it. Using one sensor with an associated logic circuit avoids this problem. Further, if more than one compressor is installed, then each compressor should be subject to a predetermined, timed, shutdown sequence, usually after idling for 20 minutes. If there is no operator attendance the start/stop initiation should be automatic.

The cyclic operation has inherent power losses at the start and end of each cycle: this gives a lower limit to the period of operation and consequently to the size of the receiver. As a guide take the minimum reservoir volume as:

$$V_{vol} = QtP/\Delta p \qquad (6.8)$$

where Δp the allowable pressure fluctuation, in bar, P the absolute pressure, t the time in seconds for the pressure to drop by Δp and Q the actual flow rate. Practice has shown that t and Δp should be greater than 30 seconds and 0.3 bar respectively.

6.6.3 *Compressor inlet valve lifter*

There are two types of valve lifters available:

(a) Devices that keep inlet valve(s) permanently open when actuated. These prevent the gas from being compressed, but allow it to enter the cylinder and be expelled again. They are mainly used on double acting cylinders. Note that on uncooled cylinders such a valve lifter can be used on one end only, usually the outer one, as otherwise the danger of overheating the cylinder exists.

(b) Lifters that prevent the inlet valve to close once compression commences and close at some stage during the compression stroke and so allow stepless operation. These are mainly used on low power machines such as small air compressors.

6.6.4 Clearance volume variation

Increasing the clearance volume is an effective way of reducing the flow, provided the stage where the clearance volume is altered has a high compression ratio. Clearance volume variation should not be used if the compression ratio is less then two. Increasing the clearance volume will slightly lower the efficiency of the cylinder.

Two systems are available:

(a) Stepwise altering the clearance usually one step only, sometimes two.
(b) Continuous variation by having an auxiliary piston to fine adjust the rate. The latter system is difficult to operate and if used on multi-stage compressors should be restricted to the first stage only.

Note that single step clearance pockets are also generally used on the first stage only. Special care must be taken when gases which may lay down deposits are handled. These tend to fall out in the dead space of the pocket.

6.6.5 Bypass

This method has the advantage of:

- quick response;
- continuous variation;
- simplest compressor construction. Further a bypass is usually needed to reduce the required torque during startup.

The disadvantage is that full load power is always required, i.e. the efficiency is lowered directly proportional as the delivery flow is reduced.

There are two principal arrangements:

(a) Bypass across the whole of the compressor
This arrangement is normally employed for starting and shutdown purposes. It covers the flow range from 0 to 100 percent. The bypass flow connection should be taken from the catchpot following the after-cooler, otherwise a bypass cooler may be required. Cooling due to the Joule Thompson effect as the pressure is reduced is only small. At high pressures and high pressure ratios let down valves are noisy and tend to wear.

(b) Bypass across first stage only
Bypassing the first stage only on multi-stage compressors reduces the energy wasted in bypassing. It allows a reduction in delivery flow by a factor about the same as the rated compression ratio of the first stage. The bypass should be

fitted between the inlets of the second cylinder(s) and the catchpot to the first stage.

6.6.6 Speed variation

Variable speed electric motors are rarely used to drive reciprocating machines. In the past turbine drives have occasionally been used. These make it possible to alter the speed and so the flowrate. However, mechanical design difficulties—caused by torsional resonances—limit such variations to less than 15 percent.

Note that reciprocating compressors operating at set pressure conditions need a constant torque whatever the speed. For this reason the use of torque convertors and fluid flywheels to lower the speed, while giving excellent control of the rate, does not result in any power saving.

6.6.7 Hybrid regulation

Most systems discussed above give step wise control. To obtain smooth control, together with low power consumption, it is common to use valve lifters or clearance volumes for coarse control and a bypass to fine trim.

6.7 Number of stages-control included

Once all the pressure variations due to rate control are fixed, check that all stages meet the delivery temperature criterion under all conditions. If not:

(a) could the pressure ratio in some of the stages be increased? or
(b) is it necessary to add an extra stage and re-evaluate staging?

In paragraph 6.3 a way of calculating the polytropic coefficient 'n' which allows a more accurate estimate of delivery temperatures is given. With small (piston diameter <200 mm), low pressure, water cooled machines the polytropic index and consequently the delivery temperature will be significantly lower than that calculated taking the polytropic index to be equal to the ratio of specific heats. For larger piston diameters and high pressures the drop in temperature due to heat transfer across the cylinder walls is small and the polytropic index will approach κ.

6.8 Configuration

Multi-cylinder machines are used to minimise forces and moments on foundations and variation in crankshaft torque and electricity supply. Machines with an even number of cranks offer a better balance than those with an odd number. This is of particular importance if the compressor is to be erected on a steel structure rather than on a solid, concrete foundation. If two or more compressors are installed close together then it is advisable to place them all onto one common foundation, so increasing the moment of inertia of the combined foundation by a factor of four or more and, in this way, reducing the amplitude of vibrations of the whole system.

Horizontal opposed cylinder configuration is the most common for large compressors (>600 kW) though the design requires more space than vertical machines. For cylinders to be self draining (on horizontal machines) inlet connections should always be on top with delivery ports at the bottom.

For oil-free machines, especially labyrinth compressors, vertical cylinders are preferred. Horizontal cylinders need a rider ring or band to support the weight of the piston. Such bands are usually fitted between the piston rings. On oil-free machines wear of the rider band is often a problem.

Compressors with a rate less than 0.5 m³/s frequently have two cylinders arranged in 'V' configuration. For higher pressure ratios a 'W' layout is often used.

Exhausters have a very large first stage cylinder which is best arranged vertically with a horizontal second stage forming an 'L' configuration.

6.9 Mean piston speed

Experience has shown that to obtain long reliable service the piston speed has to be limited. In the past speeds of 4.5 and 4 m/s for lubricated and unlubricated machines respectively were found to be the limit.

Note that though developments in materials have resulted in a continuous rise in permissible speeds, and that this development is likely to continue, there are applications which require lower piston speeds. Past experience should guide the designer.

6.10 Frame size

To select a suitable frame size start by assuming that there is only one double acting cylinder per stage. To estimate the power of a stage divide the total power P_{tot} as calculated in equation (2.2) by the number of stages E estimated in paragraph 6.7.

A first estimate of rod load R_l is

$$R_l = 2P_{tot}/Eu_p \tag{6.9}$$

where u_p the piston speed.

A more accurate rod load must be calculated once the cylinder dimensions are defined, see paragraph 6.13.

In the past compressors with rod loads of 500 kN (50 tons) were obtainable. Some manufacturers even offered rod loads up to 1000 kN (100 tons).

A rough relationship between stroke and rod load is given in Table 6.2. For better information check vendor catalogues.

Once stroke and the mean piston speed limit are fixed, the crank speed is also given. The largest machines with maximum rod loads have strokes of 500 mm,

Table 6.2   Approximate stroke and piston rod diameter against rod load

Rod load	kN	50	100	200	400	800
Stroke	mm	150	250	300	400	500
Rod diameter	mm	30	40	60	100	150

limiting the crankshaft speed to 4.5 r/s. Note that small compressors with strokes of 125 mm have maximum rod loads of 10 kN.

6.11 Cylinder design

For low and medium pressures—up to about 100 bar—double acting cylinders—handling one stage only—are preferred, see Fig. 6.5. Arrangements with more than one stage per rod are frequently used on small high pressure machines but are best avoided on larger machines. A particular problem is the design of HP cylinders. To ensure lubrication of the little end, a bearing with only a fraction of a turn before the movement reverses, the rod load must reverse, i.e. the crosshead, should be pushed as well as pulled in every stroke. This is not a problem with low pressure cylinders, where the area at the two ends of the piston is nearly equal, and the diameter of the piston rod is small compared to that of the piston, or with HP cylinders with a tailrod, see Fig. 6.5.

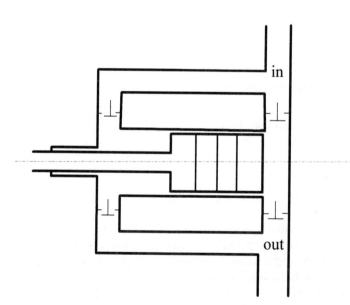

Fig 6.5   Double acting cylinder

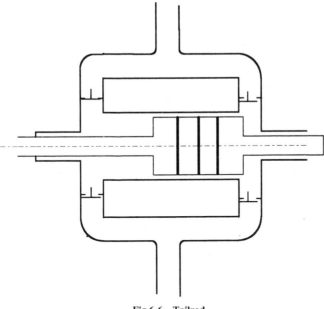

Fig 6.6   Tailrod

6.11.1   *Tailrods*

At higher pressures tailrods are needed to balance the piston forces. This solution is expensive as a complete extra packing is needed, and a 'tailrod catcher' has to be specified to guard against the tailrod being shot out in case of it breaking. Such a catcher must be capable of safely bringing the tailrod to a stop i.e. absorb all its kinetic energy if the tailrod happens to break. Tailrods are unavoidable with HP circulators (HP-stages with a pressure ratio <1.3).

6.11.2   *Single acting pistons*

An alternative for high pressure duties are single acting pistons. When operating machines with single acting pistons, special care is needed to ensure that load reversals occur. If it does not, little end failure will result. Unidirectional loading is particularly likely to occur at reduced delivery pressure, and all duties at less than design delivery pressure, including startup, should be carefully checked to ensure that load reversal occurs.

Single acting cylinders can be arranged in two ways:

(a) Working on the outer end of the cylinder and connecting the inner end permanently to the delivery is favoured for duties above 300 bar, see Fig. 6.7. The advantage is that the crossbores in the cylinder that are subjected to fatigue can be minimised; the disadvantage is that the piston rod packing sees a high and constant pressure.

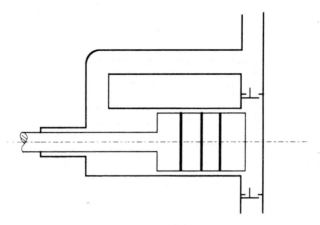

Fig 6.7   Outer acting

(b) Working on the inner end and connecting the outer end to the suction of the HP stage, see Fig. 6.8, results in the rod packing seeing a fluctuating pressure, though stresses in the cylinder are high.

Arranging two or more stages on one piston rod used to be common practice, but nowadays is restricted to small high pressure compressors.

6.12   Volumetric efficiency

The indicated volumetric efficiency η_v of a cylinder, as shown above, depends on the clearance volume ratio ε and the pressure ratio Π (equation 6.4).

This efficiency must be calculated for each stage. For a first estimate assume that clearance volume is 12 percent of the swept volume.

The actual volumetric efficiency η_{va} is somewhat lower, particularly for small machines, due to the heating and cooling effect of the cylinder walls. Allow 5–10 percent less for stages with a compression ratio less than three, more for small cylinders and higher compression ratios.

It is usually not possible to reduce the clearance volume below 10 percent as there must be some clearance at the end of the stroke (bumping clearance) and there is also some volume in the passages to and from the inlet and delivery valves. In particular a cylinder with more than one inlet/delivery valve has a number of passages; on the other hand reducing the valves to one only results in high velocities in the valves and corresponding high pressure losses.

Note that with small cylinders the deviations from adiabatic conditions inside a cylinder are greater than with large ones, due to increased heat transfer from and into the gas from the cylinder walls.

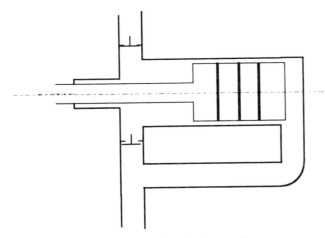

Fig 6.8   Inner acting

6.13   Piston diameter

A double acting piston without a tailrod has an area

$$A_{out} = \pi/4 D^2 \tag{6.10}$$

and

$$A_{in} = \pi/4(D^2 - d^2) \tag{6.11}$$

on the outer and inner end respectively. The total 'swept area' during a complete cycle is therefore

$$A = \pi/4(2D^2 - d^2) \tag{6.12}$$

The swept volume becomes

$$V_{swept} = Au_p/2 \tag{6.13}$$

and

$$V_{swept}\eta_{va} = Q \tag{6.14}$$

from this the required piston diameter D can be calculated to be

$$D = \sqrt{\{4Q/(\pi u_p) + d^2\}} \tag{6.15}$$

The rod loads on the outward and inward stroke R_{1out} and R_{1in}

$$R_{1out} = A_{out}p_d - A_{in}p_s \tag{6.16a}$$

and

$$R_{1in} = A_{out}p_s - A_{in}p_d \tag{6.16b}$$

Note that for single acting pistons only the inner or the outer area of the piston is active. Once the piston dimensions are known the rod load for single acting pistons can be calculated from:

	Outer acting	Inner acting	
Outward stroke	$R_{1\text{out}} = p_d d^2 \pi/4$	$(p_d d^2 - p_s D^2)\pi/4$	(6.17a,b)
Inward stroke	$R_{\text{lin}} = (p_d d^2 - p_s D^2)\pi/4$	$p_d d^2 \pi/4$	(6.18a,b)

where p_d and p_s are discharge and suction pressure of the cylinder and D and d the diameters of piston and piston rod respectively.

The two values should be of the same magnitude—one positive one negative. With the pressures obtained in paragraph 6.7 calculate the rod loads for all stages and for all operating conditions. Remember that the rod load on the last stage is a maximum at minimum throughput, whereas that on all the other stages it is a maximum at maximum throughput. If the loads are within the prescribed limits proceed—otherwise adjust the design.

In addition to the gas loads there are the mass forces due to the acceleration of the reciprocating masses. They are at a maximum at top and bottom dead centre, and have to be vectorially added to the gas forces. At high pressure ratios their effect is to lower the total rod load somewhat, at very low pressure ratios (circulators) they increase the rod load.

6.14　Number of cylinders per stage

The number of cylinders per stage is based on three quite separate considerations:

(a) Is the rod load acceptable? or should there be two or more cylinders per stage?
(b) Single stage machines: should the rate be split and two or four cylinders be used to allow a smaller frame size with better control and lower unbalance forces?
(c) Cylinder bores should be limited to about twice the stroke except for cylinders arranged in 'V' and 'L' layout where the limiting piston bore is three times the stroke. This may require the use of two first stage cylinders.

Once the stage and cylinder arrangement has been decided the power consumption should be rechecked by calculating the power requirements of each stage. For this the pressure drop between stages and changes in inlet temperature from that of the first stage should be included.

The total gas power is the sum of the power of all the individual stages. The shaft power is obtained by including the mechanical efficiency. For small compressors this is about 88 percent, for large machines it can increase to 94 percent.

$$p_{\text{shaft}} = \Sigma\, p_{\text{rot}}/\eta_{\text{mach}} \qquad (6.19)$$

6.15 Compressor valves

All reciprocating compressors have self-acting non return valves. Designs vary, plate and ring valves have nearly totally replaced poppet valves which nowadays are only used for very high pressures (500 bar+).

For valves to work economically, reliably and for a long period the gas velocity through them, as well as the lift, must be limited. The gas velocity depends on the piston velocity and the ratio of piston area to the valve area. In a valve the limiting area must be the area opened in the valve action, otherwise the valve will flutter. This area should be less than 80 percent of the narrowest fixed passage. The latter is dependent on the design of the cylinder; note that increasing valve area also leads to increasing clearance volume due to the greater number and larger size of ducts needed in the cylinder. Valve lift is limited by the compressor speed. It is in the order of 2 mm.

A guide to gas velocity u_v in valves is

$$u_v = 40 p_s^{-0.125} \rho_s^{-0.5}. \qquad (6.20)$$

If gases of different molecular weights have to be handled a compromise between pressure losses in the valves at high molecular weights and the need to prevent flutter and corresponding short valve life when handling gases of lower molecular weight should lean towards higher velocities and pressure losses than would be selected for a machine dedicated to a single duty.

Circulators ($\Pi < 1.3$) require valve velocities about 30 percent lower. The valve velocity and free valve area on submitted designs should be checked.

Fitting a suction valve in the delivery or reversing a delivery valve results in major damage to a compressor. It is therefore mandatory that valves are:

(a) not interchangeable and
(b) that it is not possible to install valves in the wrong position

The moving parts in a compressor—the plate or rings—are items that are highly loaded in fatigue. Metallic valve plates should be fully machined, not punched out of sheet; a high surface finish is also required (0.2 mm CLA). Plastic ring valves have recently become popular. Failure of valves is the main cause for random outages. It should be noted that well designed compressor valves have a long life provided that no foreign bodies or liquid droplets are trapped in the valve. Extremely high stresses occur in the valve plates if they are seating on debris or droplets. These stresses lead to rapid failure.

6.16 Bearings and lubrication

The motion work is usually lubricated with a pressurised system that supplies cooled and filtered oil to all bearings at about 2 bar. Only some small machines of less than 20 kW rely on splash lubrication. Bearings are white metal lined shells both for the main and big ends. Tilting pad bearings are not required. Little ends usually run in bronze bushes; as mentioned above it is essential that the load in the little end reverses during every turn to allow oil to enter the loaded areas of these bearings. Crosshead guides are in cast iron. The slippers—the rubbing part of the crosshead—having a white metal surface.

Shaft driven oil pumps are preferred, with an electric standby pump used for startup. Outboard motor bearings are sometimes supplied from the compressor oil system. Wet sump—with the crankcase acting as the oil reservoir—and dry sump machines are available.

Each cylinder and each gland, when lubricated, is supplied from one or more injectors on a separate total loss system. Lubricators are usually fitted with spare injectors one of which is used to operate a pressure switch to indicate that:

(a) there is oil in the reservoir of the lubricator and
(b) that the unit is working.

6.17   Piston and rider rings

To minimise bypassing between piston and cylinder, rings are set into the piston. On lubricated machines metallic rings are normally used. On dry running machines carbon or plastic is the preferred ring material.

In horizontal cylinders the weight of the piston must be supported with a 'rider-ring'. Wear of rider and piston rings usually controls planned shutdown intervals.

6.18   Piston glands

On crosshead machines, a seal is needed where the piston rod enters the cylinder. Segmental packing rings are the preferred solution.

The number of rings installed increases with increasing pressure. Vent connections and purge connections are installed in the gland packing if toxic or flammable gases are handled. With hot and HP applications the gland should be cooled.

On oxygen and other duties where it is essential that no oil is carried over from the motion work into the cylinder, the distance between the wiper on the crosshead guide and the entrance into the gland should be equal to or longer than the stroke + 25 mm, to avoid carry over of lubricants.

6.19   Crankcase

The crankcase is the backbone of a reciprocating compressor. Through its centre runs the crankshaft supported on main bearings placed between each or every second throw. Small crankcases are sufficiently rigid so as not to flex under the cyclic loads nor to deform by forces from their mountings which are often flexible. Large machines, particularly four and six throw horizontally

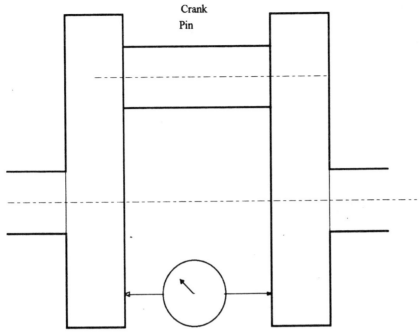

Crank
Pin

Fig 6.9   Crankshaft deflection

opposed compressors require the crankcase to be mounted on a rigid foundation to increase its stiffness. The foundation is also a base for the driver. For directly coupled motors, particularly those with an outboard bearing, great rigidity of the combined base is needed. Special care has to be taken when erecting compressors to ensure that all bearings are in line. To guarantee that all main bearings are in line and do not cause the crankshaft to flex, a final check measuring crankshaft deflections needs to be made prior to finally grouting the crankcase to the foundation (see Fig. 6.9). Crankshaft deflections are measured by measuring the gap opposite a crankpin while the crankshaft is being turned. This is done before the conrod is installed.

6.20   Materials

6.20.1   Motion work

Crankcase	Cast iron or fabricated steel
Crankshaft	Light duty Nodular cast iron
& conrods	Heavy duty Forged steel
Crossheads	Cast iron with white metal slippers
Crosshead guides	Cast iron or fabricated. Can be integral with the crankcase

6.20.2 Cylinders

The following is for guidance only; special materials may be required for corrosive or cold gases.

Cylinders	LP cast iron, nodular cast iron
	HP cast steel, forged steel
Liners	Spun grey cast iron
Pistons	LP cast iron or fabricated steel
	HP cast iron or forged integral with piston rod
Piston rods	Surface hardened steel
Piston & rider rings	
Lubricated:	Cast iron, bronze or teflon @
Non lube:	Carbon or teflon @
Packings lubricated	Bronze & cast iron

Note that parts subjected to fatigue load, such as crankcases, cylinders and pistons, if fabricated, must be free of stress raisers in the welds.

6.21   Vessels

6.21.1 Suction catchpots

For suction catchpots to act as pulsation dampers they should have a capacity of at least 30 swept volumes of one first stage cylinder. (The swept volume of one end of a double acting cylinder.)

6.21.2 Interstage vessels

The interstage vessels and pipework are usually arranged by the machine vendor.

6.21.3 Interstage and after coolers

Coolers for reciprocating machines are subjected to vibrations due to the non-continuous flow pattern. Particular attention must be paid to the design of such coolers. To avoid fretting at baffle plates, plastic ferrules are often used. Interstage coolers where condensation of water or other components takes place should be followed by separators to prevent liquid reaching the inlet valves of the next stage. Coolers on high pressure duties should have bursting discs on the water side in case of a tube fracture. Failure is more likely on such coolers as they are subjected to continuous pressure fluctuations.

6.21.4 Separators

Condensate formed should be separated and drained. Cyclone or blind separators are suitable.

Wire mesh demisters should not be used upstream of a cylinder as they tend to break up due to the pulsating flow.

Catchpots following separators should have a capacity of at least 8 hours to allow them to be emptied three times a day. Automatic drainage devices have proved to be unreliable in the past on high pressure machines (>50 bar). Separation becomes increasingly difficult with increasing pressure and condensation should be avoided at pressures above 200 bar to avoid the need for separators.

6.21.5 Pulsation dampers

Reciprocating compressors generate line pressure pulsations due to the cyclic gas velocity variations. To meet acceptable pulsation amplitudes dampers are frequently used. A simple measure of these pulsations in a uniform pipe line (a line of constant diameter) is the 'Basic Pulse' B

$$B = u_a u r \qquad (6.21)$$

where 'u_a' is the speed of sound and 'u' is the gas velocity in the pipe.

This formula applies to a system without an overlap between individual pulses. When there is some overlap with two or more double acting cylinders on one stage, the actual velocity changes (as well as the pressure pulses) become less. If the basic pulse is too large, and likely to cause problems, a pulsation damper is needed.

Examination of the above equation shows that this pulse is directly dependent on the variation in gas velocity, speed of sound and density.

Instrument air compressors tend to be connected to large receivers with capacities of 10 to 20 minutes, with a short large diameter pipe so avoiding a large pulse and any resonance.

When pulsation dampers are fitted they should be as close to cylinders as feasible to avoid any resonances in the connecting pipes. The maximum length l between cylinder and damper should be less than

$$l = u_a/25f \qquad (6.22)$$

where f the basic frequency of the pulsation in Hz

This formula can be used as a guide, though on complicated systems an analog or digital analysis should be carried out.

Note that even pulses of small amplitude can cause damage to a system if the pipe layout has an acoustic resonance. It is, therefore, advisable to carry out an analysis of the complete system to ensure resonances are avoided.

The use of large diameter pipes, such that the gas velocities are low is often favoured if light gases are being handled.

Interstage and delivery pulsation dampers should have a capacity of 10 swept volumes or more. Note that attenuation can also be achieved with smaller capacities albeit at the cost of greater pressure drops and therefore greater operating cost.

If unacceptable pulsations are experienced in the pipeline an emergency measure is to install an orifice plate of a diameter 50–60 percent of the normal delivery pipe close to the inlet or exit of an inlet and delivery vessel respectively (not close to the cylinder).

6.22 Instrumentation

The complexity of instrumentation will generally be selected by the ultimate plant operating company as befits the control, monitoring, maintenance, safety and operations policies in force. Temperature and pressure measurement should always be provided not only at the inlet and delivery of the machine but also at the inlet and delivery of each cylinder so that any malfunction can be quickly diagnosed and rectified. Seismic probes mounted on the crankcase are frequently used to warn/trip after major problems. Note that an excess of instrumentation tends to lead to spurious trips! Pressure tappings into each cylinder to enable indicator diagrams to be taken are a useful refinement, particularly for compressors operating on gases with a low ratio of specific heats.

6.23   Layout

Space management is vital at an early stage of any project. Overall dimensions should be provided and agreed, not only for the compressor itself, but also for the associated pipework, coolers and pulsation dampers.

These are some of the points that have to be addressed:

– Will coolers be underslung or mounted on the side?
– What space is required for pulsation dampers and separators?
– Is the machine to be elevated or at grade level?
– Will a mezzanine level be required?
– Will a local instrument panel be needed?
– What space is required for the withdrawal of pistons and cooler bundles?
– Where is the lay down area?
– Will there be a crane available in the compressor house or are lifting beams needed for routine maintenance?
– Can the crane lift the motor or is access needed for a mobile crane?
– Is the lubrication system integrated or separate and skid mounted?

6.24   Noise

Reciprocating compressors are relatively quiet machines. However motors, and in particular gearboxes, often generate more noise than can be tolerated in a manned compressor house (worker exposure) or in the environment. Noisy items, such as letdown valves, can be silenced with appropriate design and acoustic hoods. Compressors themselves should not be covered—they need to be attended regularly. Noise due to gas flow can be limited by keeping velocities in pipes low.

6.25 Safety

Safety considerations require that containment of the gases handled is guaranteed, especially in the case of flammable and toxic gases.

6.25.1 *Normal operation*

All stages must be prevented from over pressure. If such conditions can occur relief valves shall be fitted on interstages.

Some leakage will always be present at piston glands. Providing a vent and purge connection within each gland, so that any leakage can be piped to a safe location, is the favoured solution. This has the added advantage that the leakage rate can be measured, the best indication of wear in the gland. Sometimes the distance piece between motion work and cylinder is also purged.

6.25.2 *Failure*

Avoiding failure is the surest way of avoiding accidents. API618(4) specifies cast steel cylinders for flammable gases. Two other sources of possible failure are mixing up inlet and delivery valves (paragraph 6.16) and breaking of a tail rod (paragraph 6.11.1).

6.25.3 *Crankcase explosions*

Crankcase explosions have been frequent in the past and can cause considerable damage. Such explosions can reach a pressure of 8 bar. It is therefore practice to fit explosion relief valves with flame traps on crankcases which cannot withstand 8 bar. The opening area of the vents should be $0.07\ \text{m}^2/\text{m}^3$ of crankcase volume. Small crankcases need no vent valve provided that they have been pressure tested to 12 bar. Note that such explosions usually occur when 'blue smoke' is rising due to oil evaporating at a hot spot. The practice of having a 'quick check' to see nothing runs hot, after an overhaul, is to be avoided.

6.26 Drivers

In paragraph 6.14 above the compression power has been calculated. To this the mechanical losses must be added. Assume that the mechanical efficiency is 88 percent for medium sized machines and rises up to 94 percent for large units.

All drives should be rated to be 10 percent larger than the maximum calculated power demand.

6.26.1 Electric motors

For large machines (above 750 kW) a direct rigidly coupled synchronous motor with an outboard journal bearing are the preferred drives.

Drivers below 750 kW are usually direct rigidly coupled and can be either overhung or with an outboard bearing. Induction motors are preferred.

For ratings below 220 kW belt drives or direct coupled overhung motors should be used.

6.26.2 Steam turbines

Steam turbine drives are sometimes used if excess steam is available. Double reduction gears are required. Turbines are not recommended. However, if they are used a careful torsional analysis is required to clearly define the speed ranges where operation is possible.

6.26.3 Others

Other possible drivers are gas turbines, and diesel engines. The latter are favoured for standby machines.

Integrated compressor designs, where the compressor shares the crankcase and crankshaft with a diesel or gas engine, used to be popular in the past.

6.27 Other considerations

6.27.1 Pressure

Do not specify interstage pressures for multi-stage machines unless the process requires these to be closely defined to accommodate side branches. Allow the vendor as much freedom as possible to optimize the design.

6.27.2 Gas composition and impurities

(a) Particulate contamination

Solid carryover causes valve trouble as well as cylinder and ring wear and should therefore be avoided. The gas should not contain hard crystalline particles larger than 3 μm.

Most process gases are sufficiently clean provided that the gas has been through a condensing cooler and an effective droplet separator, but even ostensibly clear dry gases need careful evaluation. Permanent gas filters are needed where products of corrosion of preceding equipment can be carried forward or where catalyst dust may be present.

(*b*) *Droplets in suspension*

Liquid droplets significantly affect valve life. Where condensation occurs droplet separators with an effective cut off size of 10 μm should be included.

6.27.3 Polymer deposits

Gases that are likely to form polymers require a washing system to dissolve and remove such deposits.

The following should be provided:

(a) an injection system that atomises the solvent
(b) motorised continuous barring after the machine is stopped till it is purged, and
(c) an effective drain system

6.27.4 Molecular weight variations

Such variations do affect the volumetric efficiency of labyrinth compressors because piston bypass increases with decreasing molecular weight.

Further, extreme variation in molecular weight (and in operating pressure) may reduce valve life due to variation in gas forces acting on the valves.

6.27.5 Variation in the ratio of specific heats

The ratio of specific heats has a large effect on the volumetric efficiency, particularly on stages with a large clearance volume.

6.27.6 Compressibility variation

Operation near the saturation line and in particular near the critical point requires careful assessment of startup conditions and the likely fluctuations in pressure and temperature. Liquids in reciprocating compressors can cause severe damage and should therefore be avoided just like two phase flow. Use entropy charts if in existence.

When operating with the gas in the quasi-liquid condition velocities in the valves must be kept low—as appropriate with liquid pumps.

6.27.7 Absence of oxygen

Gases derived from cryogenic separation or molecular sieve plants are often devoid of oxygen—even in the form of water vapour. This prevents the oxide layer, which is essential for lubrication, to reform on e.g. cylinder walls. This in turn leads to scuffing. For low pressure applications (30 bar) the use of labyrinth compressors solves the problem. At higher pressures (>50 bar) some means of re-establishing an oxide layer must be used. Water vapour injection or periodically exposing wear surfaces to the atmosphere are possible solutions.

6.28   Operation

Most compressors have to be unloaded before they can be started. A full flow vent or bypass will be required. Standby units have to be protected from reverse flow by an external non-return valve.

Prior to finalising the line diagram an operability or hazard study should be undertaken.

Note that it is often possible to modify particularly demanding auxiliary duties in consultation with the process designers.

When selecting a compressor it has to be borne in mind that the compressor is part of a larger system into which it must be integrated.

Particular care should be taken that the gas at the first stage suction is at the design conditions. Particulate matter whether solid or liquid will cause erosion, and in the case of reciprocating compressors valve damage may cause lubrication problems. A knock out vessel can take the form of a large drum with a filter element, or can be a cyclone separator designed for the prevailing conditions. Whichever type is used any particles separated must be removed and re-entrainment must be avoided.

6.28.1   *Testing*

Reciprocating compressors are positive displacement machines not dynamic machines, hence it is not necessary to call for a full performance test. The machine will deliver the system pressure, and the flowrate is largely a function of the mechanical condition of the cylinders, etc.

Machines that operate without foundations can be tested mechanically, and performance wise in a manufacturer's works. Large machines that rely on a foundation are usually only turned over, though a complete assembly of all parts, i.e, machine, vessels, pipes and instruments on a dummy foundation is strongly recommended. Acceptance tests are difficult. Accurate flow measurement is affected by any residual pulsation.

6.28.2   *Startup*

When machines are started up for the first time, if the inlet lines have not been pickled, there invariably will be millscale and other construction debris in the inlet system, which may not have been removed during the pre-commissioning period. In the absence of a permanent filter a temporary one should be installed, designed and fitted by the vendor of the compressor. Temporary filters should be inspected frequently and removed as soon as possible—in less then 500 hours. Filters have a nasty habit of failing in service and sending all collected debris forward at one time.

6.29 Other types of reciprocating compressors

6.29.1 *Diaphragm compressors*

A special type of reciprocating compressor is the diaphragm compressor. These machines are suitable for compressing small rates ($0.03 <$ actual m^3/s) to high pressures (<1000 bar). The process gas is only in contact with metallic surfaces; the gas is never in contact with oil and therefore is completely oil free. Further these machines are absolutely leak free. Instead of the piston acting on the gas a diaphragm is flexed in a lens shaped chamber by a single acting piston pump that reciprocates in an oil-filled chamber. The oil in turn moves a diaphragm to and fro (see Fig. 6.10).

6.29.2 *Labyrinth compressors*

Another special type of oil free compressor is the labyrinth compressor. Instead of preventing gas flow past the piston with piston rings these machines have pistons which at no time touch the cylinder wall. The piston is guided from the crosshead. Sealing is achieved by having numerous grooves cut not only into the piston but also into the wall of the cylinder (see Fig. 6.11). Provided the clearance between piston and cylinder is small and the piston speed high (5–6 m/sec) good sealing and high efficiencies can be obtained.

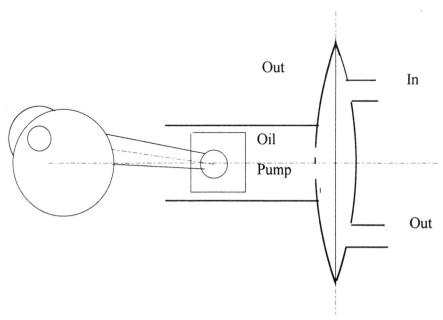

Fig 6.10 Diaphragm compressor

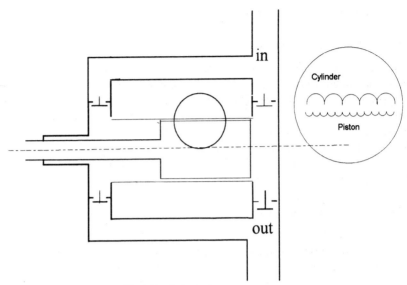

Fig 6.11 Labyrinth compressor

6.29.3 *Hypercompressors*

In the production of some chemicals, e.g. polyethylene, extremely high pressures, in the order of 3000 to 4000 bar, are required. Hypercompressors are used to raise gases to the reaction pressure. The final stage(s) of the reciprocating machines used are similar to plunger pumps: they are single acting on the outer end only. However, a single plunger attached to a crosshead would cause uniform loading on the little end bearing. To achieve load reversal it is practice to connect two opposing plungers to one crosshead (see Fig. 6.12).

6.30 Example

A lubricated compressor is required to compress 3 m³/s (10,800 Nm³/h) from 15 barg 20 C to 180 barg. Figure 2.2 shows that only a reciprocating compressor can meet this duty.

The gas is 60 percent H_2, 25 percent CO and 15 percent CO_2.
The molar mass of the mixture: $0.6 \times 2 + 0.25 \times 28 + 0.15 \times 44 = 14.8$
The ratio of specific heats κ is: $0.6 \times 1.4 + 0.25 \times 1.4 + 0.15 \times 1.3 = 1.39$
The compressibility factor z at suction: $0.6 \times 1.01 + 0.25 \times 1 + 0.15 \times 0.9 = 0.991$
Volume at suction conditions: 0.20 m³/s allow a flow margin of 5 percent
The total pressure ratio is $181/16 = 11.3$
A two stage compressor will have a compression ratio of 3.4 and an

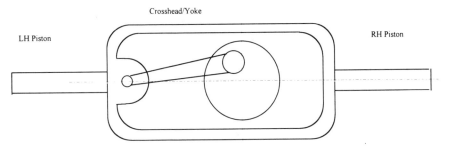

Fig 6.12 Hypercompressor

interstage pressure p_i of 53 bar. From equation (6.7) the total power is approximately:

$P = 154 \times 0.21 \times 16 \times \lg 11.3 = 1255$ kW

and from equation (6.8) the rod load R_l with a piston speed of 4.5 m/s becomes

$R_l = 2 \times 1255/2 \times 4.5 = 280$ kN

The rod diameter, from Table 6.2, is about 70 mm
Assume a clearance volume ε of 12 percent. Using equation (6.4) calculate the volumetric efficiency.

$\eta_v = 1 - 0.12(3.4^{1/1.4} - 1) = 83$ percent

assume the actual volumetric efficiency to be 5 percent lower

$\eta_{va} = 78$ percent

and using equation (6.13) the swept volume becomes:

$V_{swept} = 0.2 \times 1.05/0.78 = 0.27$ m³/s

Assuming a double acting piston equation (6.15) the piston diameter of the first stage becomes:

$D = \sqrt{\{4 \times 0.27/(\pi \times 4.5) + 0.07^2\}} = 0.285$ m

The rod load becomes: outward 242 kN inward 215 kN
This gives a good force reversal in the little end.
From equation (4.9) the temperature exit for the first cylinder is:

$$T_d = T_s\Pi^{(\kappa-1)/\kappa} = 412 \text{ K } (139 \text{ C})$$

Second Stage $p_s = 53$ bar $t_s = 25$ C

Compressibility factor: $z = 0.6 \times 1.03 + 0.25 \times 1 + 0.15 \times 0.55 = 0.95$
Assume conditions similar to those in first stage i.e. $\eta_{va} = 78$ percent
The swept volume $V_{swept} = 0.078$ m^3/s
The piston diameter $D = \sqrt{\{4 \times 0.078/(\pi 4.5) + 0.07^2} = 0.164$ m$\}$
The rod load becomes: outward 290 kN inward 200 kN
This gives a good force reversal in the little end.
The temperature exit for the second cylinder is:

$$T_d = T_s\Pi^{(\kappa-1)/\kappa} = 421 \text{ K } (148 \text{ C})$$

An addition request: efficiency turndown to 50 percent rate is required. This can not be achieved with the layout selected above. A three stage compressor is required. The overall pressure ratio remains 11.3. A horizontally opposed three cylinder compressor is badly balanced so a four throw machine is selected with two first stage cylinders in parallel. Let the first stage have a pressure ratio of 3.6 and the other two stages a compression ratio of 2 and 1.6. The interstage pressures will be 58 and 115 bara. The rod load and the swept volume of each of the first stage cylinders is half of what it was before as there are now two cylinders operating in parallel. The rod diameter is estimated to be 50 mm.

First stage $D = \sqrt{\{4 \times 0.135/\pi 4.5 + 0.05^2}\} = 0.202$ m

Second stage $D = \sqrt{\{4 \times 0.08/\pi 4.5 + 0.05^2}\} = 0.159$ m

Third stage $D = \sqrt{\{4 \times 0.04/\pi 4.5 + 0.05^2}\} = 0.118$ m

Rod loads: 1st   2nd   3rd   3rd 50 percent
in kN out   138   123   133   145 (Ignoring the lower volumetric efficiency)
in   123   90   37   97

Such a compressor would have a stroke of 0.25 m
The shaft speed would be 18 Hz (450 rpm). A directly coupled 12 pole induction motor would drive at about 490 rpm.
Conclusion: a three stage compressor running at 490 rpm would fulfil the required duty including the turndown requirement.

Chapter 7

Twin Screw Compressors

7.1 Introduction

It is assumed that you have already been through the process of preliminary selection (Part One, Chapter 2) and that a screw compressor is the appropriate choice for the duty required. There are two types of twin screw compressors: oil-free and oil-injected. Although both types use helical rotors (screws), there are many significant differences between the two types. This chapter deals with the common points; Chapters 8 and 9 deal with the special features of oil-free and oil-injected machines respectively.

Twin screw compressors consist of two intermeshing helical rotors rotating within a stator casing (Fig. 7.1). The rotor pair has protruding lobes on the male rotor meshing into corresponding flutes on the female rotor. These lobes and flutes form a multi-start helix around the rotor. There have been extensive rotor profile developments over the years and from a 'universal' profile, special shapes for different duties, such as low pressure ratio, vacuum and oil-injected, have been developed. Some commonly used shapes are shown in Fig. 7.2.

A further very important aspect of rotor profile design is the so-called 'profile combination'. The illustrations in Fig. 7.2(a) show equivalent male and

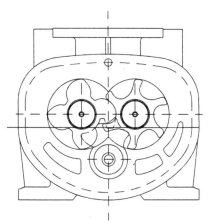

Fig 7.1   Cross-section diagram of a twin screw compressor

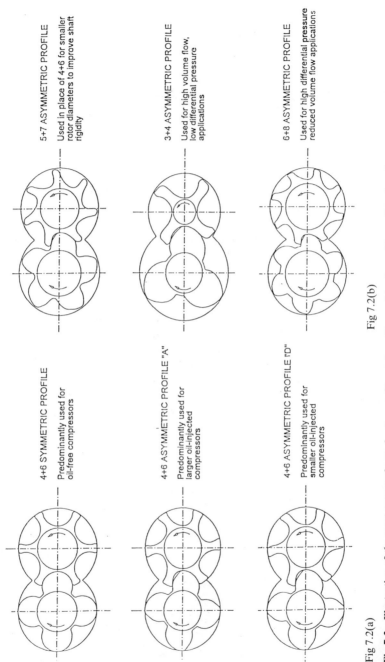

5+7 ASYMMETRIC PROFILE

Used in place of 4+6 for smaller rotor diameters to improve shaft rigidity

3+4 ASYMMETRIC PROFILE

Used for high volume flow, low differential pressure applications

6+8 ASYMMETRIC PROFILE

Used for high differential pressure reduced volume flow applications

4+6 SYMMETRIC PROFILE

Predominantly used for oil-free compressors

4+6 ASYMMETRIC PROFILE "A"

Predominantly used for larger oil-injected compressors

4+6 ASYMMETRIC PROFILE "D"

Predominantly used for smaller oil-injected compressors

Fig 7.2(a)

Fig 7.2(b)

Fig 7.2 Illustration of the most commonly used rotor profiles and profile combinations. (Note: direction of rotation assumes 'top-in', 'bottom-out'.) Ownership of all patent rights for development and manufacture of rotors with such profiles rests with Svenska Rotor Maskiner AB (SRM)

female rotor diameters, the male rotor having four lobes and the female six flutes. This is the most commonly used arrangement and is called a 4 + 6 profile combination. A smaller combination of lobes and flutes, e.g. 3 + 4, see Fig. 7.2(b), means that the volume occupied by the flutes is much larger (resulting in a higher compressor swallowing capacity), but the 'base circle' diameter of the rotor is smaller (resulting in a more slender rotor, and hence lower allowable differential pressure—see paragraph 7.4). Similarly, a higher combination such as 6 + 8, see again Fig. 7.2(b), will result in a lower volumetric capacity but higher allowable differential pressure. For the sake of simplicity, all the calculations and descriptions in this manual are based on the 4 + 6 profile combination.

7.2 Operation

The mode of operation is illustrated in Fig. 7.3. The start of the suction stroke at (a) shows gas entering the inlet port into the flutes of the rotors. As the rotors turn, more gas is taken into the flutes until at the end of the suction stroke, the suction port is sealed from the gas-filled flutes by the meshing of the next lobe. The trapped gas is then briefly moved axially along until the flutes reach the end wall (b) and the compression stroke commences. The volume trapped in the flutes is reduced until the outlet port is uncovered and (c) and the compressed gas is discharged. Note that there is no re-expansion as there is no gas trapped in a clearance volume.

The ratio of the trapped volume to the reduced volume, when the delivery port is uncovered, is the volume ratio (V_s/V_d). Note that, since this volume ratio is an inherent physical feature of a given screw compressor design, it can be described as built into the compressor and is therefore usually referred to as the 'built-in volume ratio' [V_i].

7.3 Range and capacity

Most twin screw compressors are built under licence from the Swedish company SRM who hold most of the patents referring to screw compressors. That company has developed a series of geometrically similar machines. The sizes are based on rotor diameters (D), covering an actual suction capacity from 500 to 80 000 m^3/h and 50 to 10 000 m^3/h for oil-free and oil-injected compressors respectively. An oil-free machine runs with a much higher tip speed than an oil-injected one, and therefore has a larger capacity for a given rotor diameter.

In other parts of these guidelines, methods for designing bespoke machines are outlined. However, most screw compressors are based on the SRM series and this section is therefore restricted to aid in the selection of a compressor of the right capacity. Figure 7.4 shows the favoured rotor diameters against flowrate, for an *oil-free* compressor, assuming $L/D = 1.65$ (L is length of screw rotor body), $u = 100$ m/s, profile combination = 4 + 6, $\eta_v = 100$ percent.

(a) Suction phase

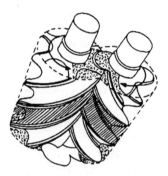

(b) Compression phase

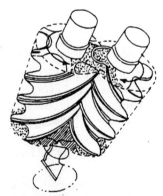

Discharge port

(c) Discharge phase

Fig 7.3   Mode of operation of a screw compressor (Mycom). (Note: direction of rotation assumes 'top-in', 'bottom-out')

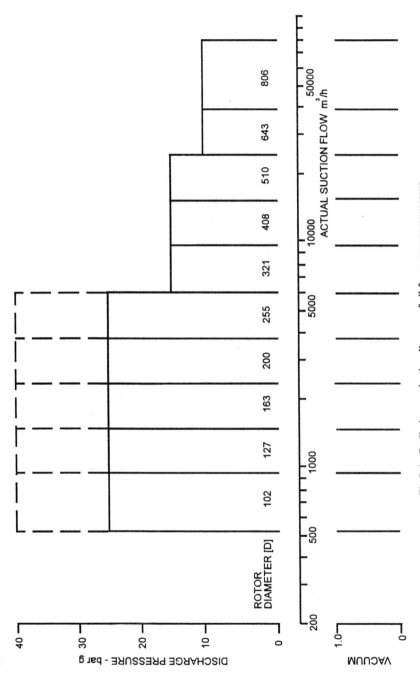

Fig 7.4 Preliminary selection diagram of oil-free screw compressors

Figure 7.5 shows similar framesizes, for an *oil-injected* compressor, assuming L/D ratios = 1.3, 1.65 and 1.93, direct-driven at 50 Hz two-pole(sizes 082 to 321) or four-pole (size 510) motor speed, profile combination = 4 + 6, η_v = 100 percent.

7.3.1 *L/D ratio*

In theory, any L/D ratio could be used. In practice a ratio of 1.65 is the most common, with various ratios between 1.1 and 2.2 also in regular use. Using a larger ratio increases the capacity at a given speed but reduces the permissible differential pressure and vice versa.

7.3.2 *Speed*

Screw compressors, particularly oil-injected ones, are slow speed machines. The normal tip speed lies around 100 m/s for oil-free, and 30–50 m/s for oil-injected machines respectively. This compares with tip speeds of between 250 and 300 m/s for centrifugal compressors. The most suitable tip speed, giving most efficient operation, depends on the molar mass of the gas handled. The volumetric efficiency (η_v) is influenced by internal gas leakage (increasing with decreasing gas density) and pressure losses (increasing with increasing molar mass), which are called the dynamic efficiency (η_{dy}). The effect on efficiency, for an oil-free compressor, is shown in Fig. 7.6; the combination of these two losses gives the internal efficiency (η_{in}). Note that the efficiency curve is flat near the maximum efficiency point; a range of speeds can therefore be used without any significant loss in efficiency.

7.4 Pressure

7.4.1 *Discharge pressure*

The discharge pressure (p_d) of screw compressors is limited by various factors—delivery temperature (t_d), L/D ratio (see below), material and casing design. Generally grey cast iron may be suitable for gauge pressures up to 15 bar. With steel or SG iron casings, pressures up to 40 or 45 bar can be achieved.

7.4.2 *Differential pressure*

A severe restriction is imposed by the permissible differential pressure (Δp). The differential pressure acts as a side load on the rotors and tends to deflect them, so reducing and opening clearances inside the machine. Hence permissible differential pressures with long, slender rotors can be as low as 2 bar, whereas short, stubby rotors can tolerate differential pressures of 25 bar and more. Figure 7.7 shows the decision-making process that indicates if a single, two- or three-stage machine is needed. It should be noted that oil-free screw

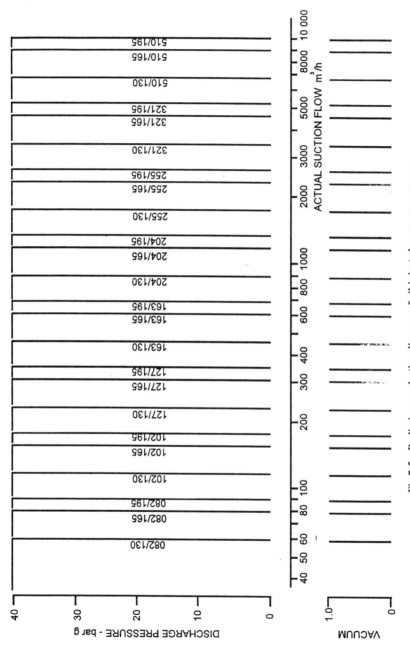

Fig 7.5   Preliminary selection diagram of oil-injected screw compressors

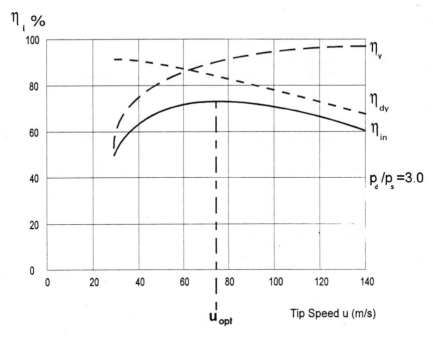

Fig 7.6   Internal efficiency (η_i) as a function of tip speed, based on oil-free screw compressor (MAN-GHH)

compressors have longer bearing spans than do oil-injected compressors (due to provision of shaft seals between rotor bodies and bearings) and, therefore, have lower differential pressure capabilities. For further details see paragraphs 8.3 and 9.3.

7.5   Volume ratio

The 'built-in volume ratio' has been defined above. Typical values are two, three or four for oil-free and up to five for oil-injected compressors. The internal pressure ratio (Π_{in}) depends not only on the 'built-in volume ratio V_i' but also on the ratio of specific heats (κ) of the gas handled. It can be calculated from

$$\Pi_{in} = (V_i)^\kappa \tag{7.1}$$

Note that efficiency drops only slightly if the actual pressure ratio differs from the internal pressure ratio (see Fig. 7.8). The extra power required, if the external pressure ratio is either smaller or greater than the internal pressure ratio, is shown by the darkened areas A and B, respectively. Note than in either case the additional power required for a 20 percent difference in pressure ratios is less then 3 percent.

The 'internal delivery pressure' is not limited by the pressure in the delivery

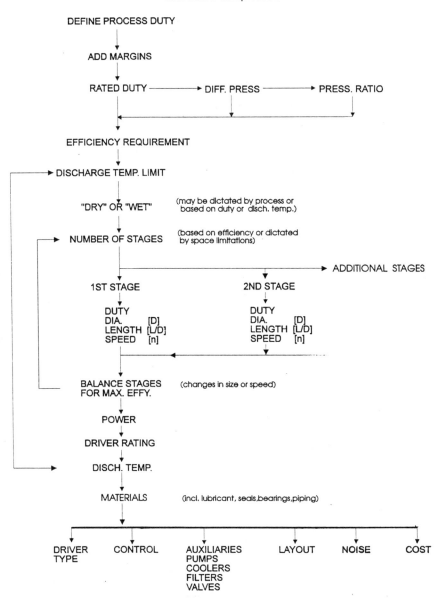

Fig 7.7   Flowchart for the selection of rotary twin screw compressors

line. The relief valve in the delivery line does not protect the compressor itself against overpressure; to prevent internal overpressures, the suction pressure of each stage must be limited to the allowable casing pressure divided by the internal pressure ratio.

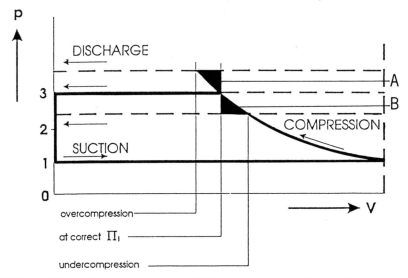

Fig 7.8 *p–V* for a twin screw compressor with $\Pi = 3.0$, operating at different discharge pressures
(p_d)

7.6 Design details

7.6.1 *Bearings*

Small machines often have rolling element bearings to take both journal and thrust loads. Larger machines have plain bearings though some retain rolling element bearings for thrust loads. Larger machines are usually equipped with tilting-pad thrust bearings, although sometimes tapered-land bearings are used. Note that tilting-pad journal bearings have no advantage in slow running machines. Plain bearings should be white metal lined shells. The choice of the white metal, tin or lead rich, is influenced by the gas handled.

Since the highest loads in a screw compressor are axial, the thrust bearing is the component most susceptible to premature failure. The compressor should therefore be protected by embedded temperature sensors, mounted in the thrust bearing pads of all the axial bearings.

7.6.2 *Seals*

Seals are required to prevent the loss of gas from the machine. In oil-free machines both rotors require a seal at each end (four seals in all) to prevent the gas from entering the bearing oil space. In oil-injected machines the oil is in intimate contact with the gas and these machines require only one seal where the drive shaft enters the casing, although there are restrictions at each end of each rotor. The types of seals used are:

(a) *Labyrinth type*
Labyrinth seals are only flow restrictors, not positive seals. When used on machines handling toxic or flammable gases they require an auxiliary system to pump the leaked gas away.

(b) *Restrictive ring type*
Restrictive ring seals, usually made in segments of a self-lubricating material such as carbon, reduce the leakage path even more than labyrinths do, but like labyrinths they work by passing some gas. They too therefore need a system to pump away any leakage.

(c) *Mechanical seals*
Mechanical seals are the most commonly used seals on screw compressors for process gases. They should be provided with labyrinths and slingers to minimise leakage. Until recently, mechanical seals only worked when the sealing faces separated gas from liquid, and some liquid, usually oil, was therefore required. Lube oil is always available and is the preferred liquid. The use of a separate seal liquid is extremely difficult. To seal such a liquid from the process gas and from the lube oil requires numerous secondary seals. Separate seal fluids are therefore best avoided.

7.6.3 *Materials*

Major items are normally supplied in the following materials:

(a) *Casings*

- For air and inert gases: grey cast iron.
- For hydrocarbons and other hazardous gases where site conditions will allow a less expensive material than steel: cast SG (nodular) iron.
- For hydrocarbon and other hazardous gases, including sour and wet gases, cast carbon steel can be used on oil-injected machines. Oil-free units must be built of high alloy steels that can withstand the corrosive atmosphere, particularly if water injection is included.

(b) *Rotors*

- Forged carbon steel for all non-corrosive applications.
- High alloy steels (13 percent Cr, 4 percent Ni) for corrosive applications; for oil-free machines, it may be necessary to forge the rotors in stainless steel (18 percent Cr, 8 percent Ni). Note that all stainless steels are prone to galling, particularly likely with screw compressors that have very fine operating clearances.

(c) *Timing gears (necessary on oil-free compressors)*

- Forged allow steel suitably hardened.

7.6.4 Lubrication system

Lube oil, seal oil and/or control oil systems should be supplied to a recognised standard, e.g. API 614 (**19**). The system should supply oil to the compressor as well as to the gearbox, coupling and motor if these are oil-lubricated. When handling some gases, especially those that go into solution in the lubricant, a separate oil unit for the compressor may be required.

7.6.5 Rotor dynamics

Screw compressors are slow-running machines with short and stiff shafts, particularly oil-injected ones. Longer shafts are needed for oil-free machines as there are seals on each end of the compression chamber. Most screw compressors operate below their first critical speed in bending. Problems may arise only with excessive overhangs and the first critical should be calculated.

7.6.6 Torsional response

It is usually necessary to verify the torsional response of the complete machine train by means of a torsional analysis that considers all the rotating masses and stiffnesses. The first and the second one on three element systems are very low and below the excitations in the system (lobe passing and gear contact frequency as well as electric excitation). If an excitation is likely the system can be detuned by modifying the coupling design.

7.6.7 Condition monitoring

Screw compressors are slow-running, insensitive machines. Performance monitoring by measuring flow as well as suction and delivery condition should be provided. However, provided the gas is not too contaminated and the oil supply is maintained (and monitored), no other monitoring on line is required. In particular, the use of shaft vibration transducers, nowadays always installed on high speed process machines, is neither beneficial nor cost-effective. If vibrations cause concern, a simple seismic transducer should be fitted to the casing. Note, however, that the provision of temperature measurement on axial bearings *is* recommended (see paragraph 7.6.1). A detailed description of screw compressors is given in reference (**20**).

Chapter 8

Oil-Free Twin Screw Compressors

8.1 Introduction

A general description of twin screw compressors is given in Chapter 7. This chapter deals with the points that apply to *oil-free* compressors only.

The description 'oil-free' implies that the gas compression space is entirely free from oil contamination, although usually the bearings and seals of such compressors are supplied with oil under pressure. The alternative description of 'dry running' is inaccurate since the gas compression space is often injected with water or a liquid solvent to partially absorb the heat of compression, to prevent premature polymerisation, or simply to wash away any particulate entrained in the gas stream.

8.2 Application

Oil-free screw compressors are ideal for a variety of gases requiring relatively low pressure ratios at fairly constant volume flow. They are particularly suitable for handling contaminated, particle-laden, polymerising or explosive gases. Typical gases include: methane, ethylene/ethane, propylene/propane, acetylene, butadiene, flare gas, strip gas, coke-oven gas, lime kiln gas (soda ash plants), platform recycle gas, visbreaker off-gas, as well as steam for mechanical vapour recompression (MVR).

8.3 Staging

Once suction and delivery pressure are known, the differential pressure and pressure ratio can be calculated and, knowing these, the number of stages required can be calculated. As a first approximation, assume equal pressure ratios across the stages, and calculate intermediate pressure(s) (p_{int})

$$p_{int} = p_s \sqrt{\{p_d/p_s\}}, \text{ or cube root for a three-stage machine, etc.} \quad (8.1)$$

Check that the differential pressure (Δp) on any stage does not exceed the maximum allowable values for oil-free screw compressors; depending on the compressor design, this can vary from as much as 15 bar for $L/D = 1.1$ to as low as 3 bar for $L/D = 2.2$. The reason for this limitation is rotor deflection: the larger the L/D value, the greater the bearing span and the larger the rotor deflection. Oil-free screw compressors have relatively small rotor clearances and permissible deflection is very small, hence the severe design limitation on differential pressure.

It should also be recognised that the differential pressure ratings given above apply to screw compressors with the conventional 4 + 6 profile combination; the use of higher rigidity rotors with, for instance, a 6 + 8 profile combination and L/D of 1.0 can provide a differential pressure capability of as much as 25 bar.

For compressor applications with sidestreams or offtakes, multi-staging is necessary. Further intercooling is often required and may result in liquid knockout. If there are no process demands for specific interstage pressures, the compressor vendor should be allowed to select these.

8.4 Clearances

In an oil-free compressor, synchronous running between male and female rotors, and consequently the running clearance between them is maintained with timing gears. As stated above, the clearances themselves, between both rotors and casing, are extremely small to allow the unit to operate at optimum efficiency, although they must always allow for thermal expansion. Figure 8.1 shows the efficiency depending on running speed and clearance. It shows that an increase of as little as 0.01 mm can lower the efficiency by as much as 1 percent.

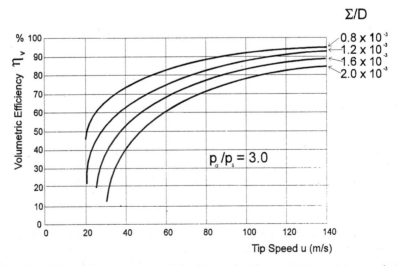

Fig 8.1   Volumetric efficiency (η_v) as a function of tip speed (u) for several internal clearance/rotor diameter (Σ/D) ratios. (MAN-GHH)

8.5 Discharge temperature (t_d)

The maximum temperature that gases are allowed to reach differs for each gas (see Table 6.1). The discharge temperature of oil-free screw compressors is further limited by the effect of thermal growth on the internal clearances of such compressors. Normally the maximum temperature rise permitted is 180 C. However, with special cooling arrangements, such as cooled rotors, temperatures up to 230 C can be handled. If gas conditions permit, the injection of a liquid (usually water) into the compression space allows the discharge temperature to be kept low.

8.6 Selecting the right size

Most oil-free screw compressors are driven through a gearbox. This may be integral in the compressor casing, or, for larger machines, be a self-contained gearbox. By selecting suitable ratios the exact required flowrate can be achieved. To calculate the optimum speed and the size of compressor required proceed as follows:

(a) The optimum tip speed (u_{opt}) is at a Mach number of about 0.3. The sonic velocity (u_a) for a particular gas is shown in Appendix II (II.2).

$$u_{opt} = 0.3u_a \qquad (8.2)$$

Note that the highest permissible tip speed is 150 m/s.
(b) Estimate the volumetric efficiency (η_v) from Fig. 8.2.
(c) Adjust flowrate Q to give flowrate at 100 m/s tip speed

$$Q_{100} = 100Q/(\eta_v \times u_{opt}) \qquad (8.3)$$

(d) From Fig. 7.5, select the next larger compressor size (Q_c) with rotor diameter (D).

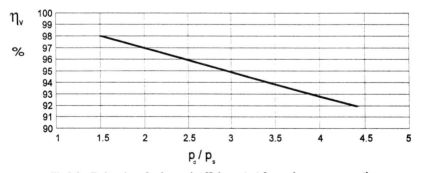

Fig 8.2 Estimation of volumetric efficiency (η_v) for various pressure ratios

(e) Calculate operating speed (N) from:

$$N = (Q_{100}/Q_c)31\,000/D \tag{8.4}$$

where N is in r/s and D is in mm

8.7 Control requirements

8.7.1 *Pressure control*

Screw compressors, like all positive displacement machines, will deliver at the pressure existing in the delivery line.

8.7.2 *Capacity control*

Oil-free screw compressors are not suited to capacity regulation, except by speed variation (see paragraph 8.7.3), since their characteristic is very steep (see Fig. 8.3).

Small reductions in the mass flow can be achieved by suction throttling provided the allowable differential pressure is not exceeded. The only other method is gas recycling, which is an inefficient method of control.

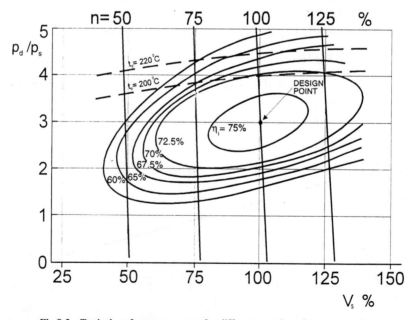

Fig 8.3   Typical performance curves for different speeds and pressure ratios

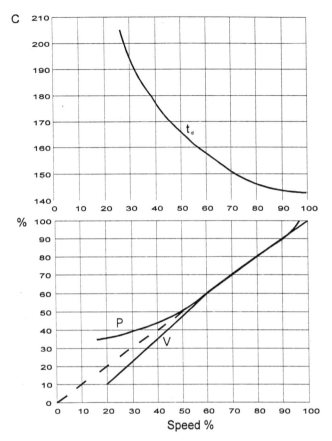

Fig 8.4 Curves of capacity (*V*), power (*P*), and discharge temperature (t_d) against speed (MAN-GHH)

8.7.3   *Speed control*

The nearly flat efficiency/speed curve near the design point shows that speed control is favoured for oil-free screw compressors. However, it must be noted that screw compressors require a constant torque which means that the anticipated power saving can only be achieved with the more sophisticated (and expensive) electric drives or with steam turbine drivers. Note also that the rise in discharge temperature is very significant at lower part-load flowrates (see Fig. 8.4) and this feature usually provides the limitation to possible turndown.

8.8   Power

It is customary to calculate the power requirements of screw compressors based on the adiabatic index (κ) and not the polytropic index (n) used for calculating dynamic machines. The adiabatic power (P_{ad}) for a given pressure ratio p_d/p_s is:

$$P_{ad} = p_s V_s \kappa/(\kappa - 1)\{(p_d/p_s)^{(\kappa-1)/\kappa} - 1\} \tag{8.5}$$

The internal efficiency η_i is about 0.7 (see Fig. 7.7). This efficiency is based on internal bypass and gas friction losses. Further, the mechanical efficiency η_{mech} is about 95 percent so that the required shaft power (P_s) becomes:

$$P_s = P_{ad}/(\eta_{in}\eta_{mech}) \tag{8.6}$$

Note that the mechanical losses do not increase the gas temperature but warm up the lube/seal oil.

8.9 Temperature

The temperature rise for adiabatic compression is given by:

$$\Delta t_{ad} = T_s(p_d/p_s)^{(\kappa-1)/\kappa} \tag{8.7}$$

The actual rise is ΔT_a; this can be calculated using the adiabatic efficiency (η_{ad}):

$$\Delta T_a = \Delta t_{ad}/\eta_{ad} \tag{8.8}$$

If the discharge temperature is known, the adiabatic efficiency (η_{ad}) can be calculated. In the case of liquid injection the heat capacity of the fluid inside the compression chamber is greatly increased and the final temperature correspondingly lower.

8.10 Noise

Oil-free screw compressors are relatively noisy machines due to the non-continuous gas delivery. Peak amplitude occurs at lobe passing frequency, i.e rotational speed of rotor multiplied by the number of lobes on the rotor. Note that male and female rotors usually have different rotational speeds and number of lobes.

8.11 Noise attenuation

Noise is emitted to the surroundings from:

(a) the machine casing: this can be attenuated with acoustically lagged enclosures and
(b) the pipework: since almost all screw compressors operate at medium speed ($N > 15$ r/s) absorption silencers are usually installed in both the suction and delivery line. Alternatively lagging the suction and delivery pipework will give some attenuation.

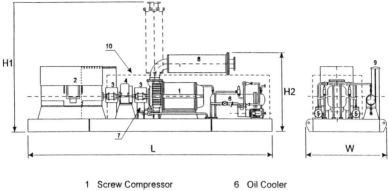

1	Screw Compressor		6	Oil Cooler	
2	Electric Motor		7	Oil Filter	
3	Spacer Couplings		8	Suction Silencer	
4	Gear Box		9	Control Panel	
5	Oil Pumps		10	Acoustic Enclosure	

Compressor Framesize	Overall Module Dimensions				Module Weight kg
	L mm	W mm	H1 mm	H2 mm	
102	3000	1500	2500	1600	7000
127	3500	1500	2800	1800	8000
163	4000	1750	3100	2000	9000
204	4500	2000	3500	2300	10000
255	5000	2250	4000	2600	12000
321	5700	2500	4500	2900	15000
408	6400	2750	5100	3300	19000
510	7200	3000	5700	3700	27000
643	8100	3250	6400	4200	42000
806	9000	3500	7200	4700	70000

Fig 8.5 General arrangement with dimensions and weights of an oil-free screw compressor module. (Note: for guidance only; precise dimensions and weights can vary considerably depending on specification requirements and scope of supply)

8.12 Layout

Size and weight of the compressor package will depend on several factors such as design pressures, material selections and purchaser's specifications in terms of scope of supply and accessibility. As a very rough guide, the general arrangement in Fig. 8.5, with dimensional matrix may be used. It should be noted that the dimensions given assume a single-stage compressor mounted on a single baseframe with integrated oil console and silencers mounted above the compressor; different arrangements will greatly affect the dimensions given.

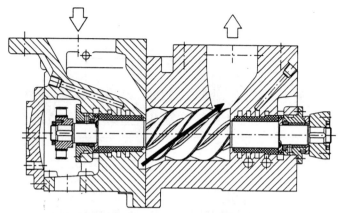

Fig 8.6(a)   Direction of flow of gas through rotors. 'Top-in', 'top-out' design; 'upward' gas flow.
(Typical, based on MAN-GHH)

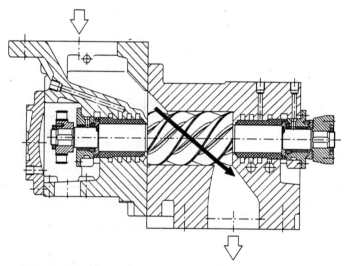

Fig 8.6(b)   Direction of flow of gas through rotors. 'Top-in', 'bottom-out' design; 'downward' gas
flow (MAN-GHH)

Nozzle orientation (see Fig. 8.6) for this type of compressor is either:

(a) top-in, top-out, which is suitable for most gases and helps to reduce the
overall installation size, or

(b) top-in, bottom-out, which is ideal for gases with large entrained volumes of
liquid, but which, in the larger frame sizes, will require a mezzanine-type
foundation.

8.13 Gas contamination

8.13.1 *Particulate contamination*

Some gases (e.g. coke-oven gas) contain hard particulate matter entrained in the gas stream. In such applications, to resist abrasive erosion, the gas should not contain particles larger than $50\,\mu$m. Usually the gas is at least partly cleaned at source, although it may be necessary to install a separator to remove particles of larger dimensions.

Other gases, (e.g. lime kiln gas in a soda ash plant) contain soft particles entrained in the gas; unless corrosive, this type presents no problem to oil-free screw compressors. The continuous meshing of the rotors not only prevents heavy buildup and ensures an even distribution of the deposited material, but the deposit also reduces the internal clearances and so increases the volumetric efficiency.

8.13.2 *Droplets in suspension*

The presence of liquid droplets above $25\,\mu$m will cause erosion. To avoid this, a separator should be installed upstream of the compressor. Further, when liquid is injected into the machine it is essential that the liquid is atomised to avoid erosion problems.

8.13.3 *Polymer deposits*

Liquid injection can help to reduce polymer deposition by keeping the compression temperature below the polymerisation temperature. Injecting a solvent while barring can help to remove deposits once they have occurred. To do this a barring device as well as a self-draining machine is needed.

Chapter 9

Oil-Injected Twin Screw Compressors

9.1 Introduction

A general description of screw compressors is given in Chapter 7. This chapter deals with the points that apply to oil-injected compressors only.

The description 'oil-injected' implies that lubricating oil or a suitable synthetic lubricant (see paragraph 9.9) is deliberately injected into the gas compression space and is allowed to mix freely with the process gas. There are five important benefits to be derived from the injection of oil into the compressor.

(a) The primary purpose of oil-injection in screw compressors is that the oil acts as a *coolant* thus enabling the discharge temperature to be closely controlled. Differential expansion due to thermal growth can result in reduction of running clearances in screw compressors, leading to interference between casing and rotors. By injecting a calculated flow of oil, the discharge temperature is controlled in the range 80–100 C (compared to 180–220 C in an oil-free compressor). This feature permits relatively large pressure rises across the compressor and obviates the need for jacket and/or rotor cooling.

(b) For optimum efficiency, internal clearances within the compressor must be kept as small as possible; however, due to the presence of large quantities of oil within the machine, there is no danger of contact between the rotors and the casing. The oil also serves as a *sealant*, partially filling the clearance between rotors, rotor-and-casing and the 'blowhole'. This results, either in an improvement in efficiency, or a relaxation of the tolerances on the Σ/D ratio chosen for the compressor, which can lead to a reduction in manufacturing costs.

(c) The injected oil acts as a *lubricant* enabling the main rotor (usually the male) to drive the other rotor through a film of oil. Thus, an oil-injected screw compressor does not require a timing gear between the rotors to ensure that there is no metal-to-metal contact. This results in a simpler, less expensive compressor for the same duty. The lubrication effect also permits the use of a slide-valve capacity control system, capable of controlling volume flow between 10 and 100 percent, providing the most flexible compression system available.

(d) Although not an initial reason for using oil-injection in screw compressors,

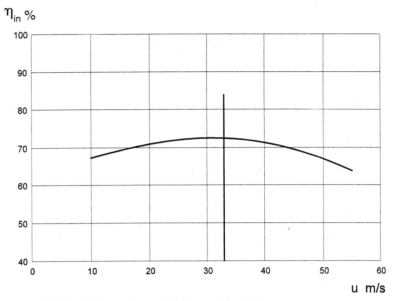

Fig 9.1   Optimum tip speed (u) for an oil-injected screw compressor

experience has shown that the presence of so much oil in the compression chamber, coating all metal surfaces, acts as a significant barrier to corrosive attack by elements of the compressed gas. The injected oil thus acts as a *corrosion inhibitor*, and this feature has permitted this type of screw compressor to be used with gases containing sulphides and chlorides, for instance, while using conventional carbon steel or grey cast iron as materials of construction. (e) A final feature of oil-injection in screw compressors is that the oil acts as a *noise-damper*; combined with low rotor lobe passing frequencies, this has the advantage that most installations do not require extensive sound attenuation to meet acceptable site noise levels.

Figure 9.1 shows how the internal efficiency depends on the tip speed. Note that the optimum tip speed is around 30 m/s, about one-third of that for oil-free machines.

9.2   Application

Oil-injected screw compressors are ideal for a variety of gases requiring relatively high pressure ratios and/or widely varying volume flows, where:

(a) an available lubricant is compatible with the gas, and
(b) an occasional very small carryover of the lubricant can be tolerated.

Typical applications include refrigerants (HFC, HCFC and CFC), ammonia, carbon monoxide, carbon dioxide, natural gas, methane, ethylene/ethane, propylene/propane, butane, hydrogen, helium, chlorine, hydrogen chloride, methyl chloride, landfill gas, digester gas, reducer gas, air and many others. The presence of substantial quantities of hydrogen sulphide and/or water vapour in the gas is usually not a problem for this type of compressor provided a suitable lubricant is chosen.

9.3 Staging

Once suction and delivery pressure are known, the differential pressure and pressure ratio can be calculated; the number of stages can then also be calculated. Note that most duties suitable for oil-injected screw compressors can be handled in a single stage of compression. For pressure ratios between 10 and 25, however, a better efficiency will be obtained from using a two-stage, or a compound two-stage, screw compressor.

As a first approximation, assume equal pressure ratios across the stages and calculate intermediate pressure (p_{int}):

$$p_{int} = p_s \sqrt{\{p_d/p_s\}} \tag{9.1}$$

Check that the differential pressure (Δp) on any stage does not exceed the maximum allowable values for oil-injected screw compressors; depending on the compressor design, this can vary from as much as 30 bar for $L/D = 1.1$ to as low as 8 bar for $L/D = 2.2$. The reason for this limitation is rotor deflection; the larger the L/D value, the greater the bearing span and the larger the rotor deflection. Although oil-injected screw compressors have larger rotor clearances and shorter bearing spans than do oil-free screw compressors, the permissible deflection is still quite small and imposes these limits on allowable differential pressure.

It should also be recognised that the differential pressure ratings given above apply to screw compressors with the conventional 4 + 6 profile combination (see paragraph 7.3). The use of higher rigidity rotors with, for instance, a 6 + 8 profile combination (see Fig. 7.3(b)) and $L/D = 1.0$, can provide a differential pressure capability of as much as 50 bar.

For compressor applications with sidestreams or offtakes, multi-staging is necessary. Intercooling between stages is not normally required in oil-injected screw compressors. If there are no process demands for specific interstage pressures, the compressor vendor should be allowed to select these. This is especially important with compound, two-stage compressors.

9.4 Clearances

As already explained in paragraph 9.1(c) the injected oil is used as a lubricant. Consequently the male rotor usually drives the female rotor (although the other way around is known) through a thin film of oil. There is, therefore, no need of timing gears to ensure synchronous running with this type of compressor. Although the running clearances between both rotors and casing are larger than in the oil-free type, they are still small enough to allow the unit to operate at optimum efficiency. Note, however, that the action of the injected oil as a sealant has the effect of closing the leakage paths to back-flowing gas and consequently oil-injected screw compressors can run at very much lower tip speeds than oil-free compressors and yet achieve the same resistance to 'slippage'.

9.5 Discharge temperature (t_d)

The maximum temperatures that gases are allowed to reach in an oil-injected screw compressor are usually determined by the lubricant. For mineral oils, the normal permitted maximum is in the range 90–100 C; some synthetic lubricants will permit this discharge temperature to be relaxed up to 130 C, although the design of the compressor will need to be carefully checked to ascertain the effect of thermal growth on the internal clearances of the compressor. Note that in this type of compressor the discharge temperature is not determined by the isentropic exponent (κ) of the gas. The quantity of oil injected into the compressor is adjusted to make sure that the combined gas/oil discharge temperature does not exceed the above values.

Another important design criterion for oil-injected screw compressors is the calculated dewpoint of the gas mixture at discharge; this is a particularly important consideration for saturated gases, or gases with a high proportion of heavy hydrocarbons. Note that the calculated compressor discharge temperature *must* exceed the calculated dewpoint temperature of the gas mixture by 12–15 C. In the case of saturated gases, reduction of suction temperature by pre-cooling and separating condensibles can bring the discharge temperature within the range permissible.

9.6 Selecting the right size

9.6.1 *Estimation of actual volume flow*

Once the appropriate rotor diameter (D) and length/diameter ratio (L/D) have been established from the framesizes shown in Fig. 7.6, an approximation of the actual gas suction volume flow (Q_s) rating of the compressor (assuming asymmetric profile 'A' and 'profile combination' of 4 + 6) can be estimated from the following empirically-derived formula (D in m; N in r/s; p_d in *bar a*):

$$Q_s = 8.4\{195 - 0.1p_d[(p_d/p_s) - 1] - (p_d/p_s)\}D^3(L/D)N \text{ m}^3/\text{s} \qquad (9.2)$$

This approximate value for suction volume flow will, in practice, be further influenced by gas density, solubility of lubricant, lubricant quantity, lubricant temperature and lubricant viscosity.

9.6.2 *Selection of compressor speed*

As mentioned in paragraph 7.3, oil-injected screw compressors are almost invariably driven by direct-coupled electric motors operating on either 50 Hz or

60 Hz systems, depending on geographical location. For framesizes 082 to 321 the male rotor speeds are 50 and 60 r/s respectively and for framesizes 510, male rotor speeds are 25 and 30 r/s, respectively.

9.6.3   *Estimation of compressor power*

Another empirical formula, subject to the same criteria as that given in paragraph 9.6.1, has been developed for the estimation of compressor shaft power (P) for oil-injected screw compressors (D in m; N in r/s; p_d in *bar a*)

$$P = 0.6\{62(p_d + 2) - [22.4(p_d + 3)(p_d/p_s)^{0.167}]\}D^3(L/D)N \ [kW] \qquad (9.3)$$

This formula also takes account of all normal *mechanical* losses.

9.6.4   *Estimation of discharge temperature*

Discharge gas temperature can normally be controlled by varying the quantity of injected oil, and is almost always below 100 C; thus, an aftercooler is rarely required.

9.7   Control

Oil-injected screw compressors are usually fitted with a slide valve that permits the suction volume to be reduced; a reduction down to 10 percent is possible. This stepless control of capacity, while maintaining a high efficiency, is one of the main advantages of oil-injected compressors. Reducing the suction volume also reduces the inbuilt volume ratio and so the internal pressure ratio. Operation away from the internal pressure ratio, as shown in Chapter 7, does not significantly influence the efficiency. Figure 9.2 shows the location of the slide valve in an oil-injected screw compressor, Figure 9.3 its operating effect diagrammatically, and Figure 9.4 plots power demand against percentage flow.

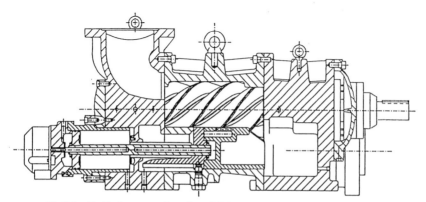

Fig 9.2   Vertical cross-section of an oil-injected screw compressor (Mycom)

SUCTION

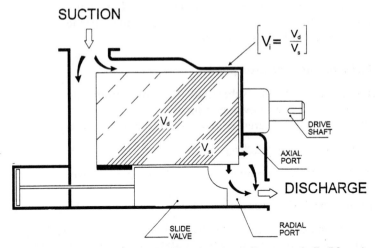

Fig 9.3   Operating effect of the slide valve, shown diagrammatically (Mycom)

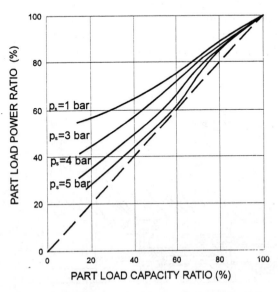

Fig 9.4   Effect of capacity control slide-valve operation (Mycom)

9.8   Economiser

Oil-injected screw compressors were originally developed for refrigeration duties. The efficiency of a refrigeration cycle is considerably improved with multi-stage expansion, provided that the gas formed in each expansion stage can be recompressed to the condensation pressure. The quantity is small compared to the total flow and, by placing a port in a position in the casing at a point where the main suction has been closed and the gas is already partly compressed, such flash gas can be recompressed.

9.9   Mechanical details

In oil-injected screw compressors the same oil is used for lubrication and for injection. It is, therefore, unnecessary to install a seal on each end of both rotors. A single mechanical seal at the inlet of the drive shaft is sufficient. Further, because of the absence of seals on the rotors these are shorter and stiffer, so allowing higher differential pressures to be applied. Nor are lateral vibrations a problem with such stiff slow running rotors.

9.10   Compound two-stage compressors

For use where multi-staging is either necessary or desirable, some manufacturers have developed a compound two-stage compressor which incorporates two standard screw-compressor framesizes integrated back-to-back in a single unit. The drive-through arrangement from the LP to HP casing enables a single main driver to be used in a two-stage configuration. The close-coupling of the two stages also allows the compressor to be packaged into a relatively small module. The gas is piped externally from the LP discharge to HP suction; there is no requirement for intercooling or interstage liquid knockout. A sectional arrangement drawing of the same compressor is shown in Fig. 9.5.

9.11   Contamination

The internal clearances in oil-injected screw compressors are somewhat larger than in oil-free machines. Therefore they are capable of digesting larger entrained particles, up to 200 μm, provided they are hard and crystalline. Soft

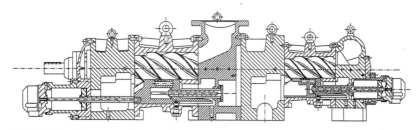

Fig 9.5   Vertical cross-section drawing of two-stage compound screw compressor (Mycom)

particles tend to form sludges with the oil and block passages. Liquid carryover can cause erosion at the inlet end of the rotors and more seriously can be trapped and dilute or emulsify oil. Separators upstream of the compressor are needed if such liquids are present in the gas stream.

9.12 Noise

Due to the sound absorbing characteristic of the injected oil, and its relatively slow running speed, an oil-injected screw compressor is much less noisy than an oil-free machine. Typical noise levels in $dB(A)$ at 1 m distance range from 80 for a 082 framesize to 88 for a 321 size. Due to the amount of oil present in the gas stream, in-line, gas-borne noise is relatively low and silencers are not required in the vast majority of cases.

9.13 Compressor module layout

The compressor manufacturer's scope of supply normally includes the complete drive train, including electric motor (or other driver), gearbox (where required) and coupling(s). Nozzle orientation for this type of compressor is almost always top-in, bottom-out, which is ideal for gases with large entrained volumes of liquid; in the larger frame-sizes, the outlet flange may be located on the side of the lower part of the casing to avoid the need for a mezzanine-type foundation design. An oil recovery/supply vessel, known as the primary separator, removes most of the oil (down to a residual oil carryover of 100–200 ppmv) and also acts as a reservoir to provide lubricating, seal and/or control oil to the compressor train. A suction strainer is provided, usually mounted directly onto the compressor inlet connection, to prevent large particulate matter from entering the compressor. The typical general arrangement drawing (Fig. 9.6), together with its dimensional matrix may be used for rough sizing of oil-injected screw compressor modules.

For applications where available space is limited, the compressor itself is small, or a lower-cost package is desired, it is common to arrange the primary separator vessel horizontally and to mount the compressor, main driver and oil management system components directly onto the separator vessel. A typical general arrangement drawing using this layout is shown in Fig. 9.7 and, together with the dimensional matrix, this may be used for rough sizing of oil-injected screw compressor modules constructed in this manner. Note that although size 321 is included in this matrix, the resultant packaged module is usually too large and heavy to utilise the horizontal separator design.

9.14 Compressor module scope of supply

The major components of the scope of supply are shown in the typical P and I diagram, Fig. 9.8, and all auxiliary equipment must comply with process industry specification requirements, such as API.

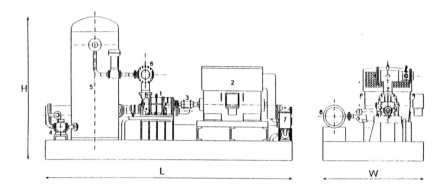

1 Screw Compressor   5 Primary/Secondary Separator
2 Electric Motor   6 Suction Strainer
3 Spacer Coupling   7 Oil Filters
4 Main Oil Pump   8 Oil Cooler

Compressor Framesize	Overall Module Dimensions			Module weight
	L mm	W mm	H mm	kg
163	4500	2500	2750	10000
204	5500	3000	3000	15000
255	7000	3500	3250	20000
321	9000	4000	3500	25000
408	10500	4000	3750	35000
510	12000	4000	4000	45000

Fig 9.6   Typical general arrangement with dimensions and weights of an oil-injected screw
compressor module with vertical oil separator

The screw compressor module has to be fully integrated into the process
system. Special care must be taken to ensure that the compressor receives gas
at the design conditions and quality. If particulate contamination of the gas,
solid or liquid, exists in large quantity, erosion and lubrication problems will
result. In these cases it is necessary to install a knock-out drum upstream of the
compressor system inlet, which may be in the vendor's or the purchaser's scope
of supply. As previously stated, the compressor is normally provided with a
replaceable suction strainer and this is augmented by a finer mesh temporary
strainer during the compressor unit commissioning period. Strainers are
usually included in the compressor vendor's scope of supply.

Non-return valves (check valves), as well as isolating valves, are usually
installed at both inlet and/or outlet of the compressor module; the NRV at
outlet should be located as close as possible to the compressor discharge flange

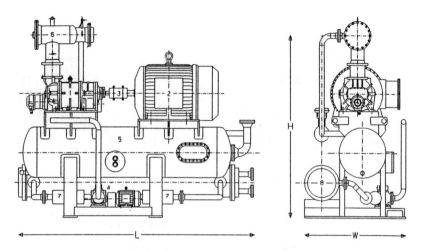

1 Screw Compressor	5 Primary/Secondary Separator
2 Electric Motor	6 Suction Strainer
3 Spacer Coupling	7 Oil Filters
4 Main Oil Pump	8 Oil Cooler

Compressor Framesize	Overall Module Dimensions			Module weight kg
	L mm	W mm	H mm	
082	2500	1000	1800	3000
102	2800	1000	1900	4000
127	3000	1100	2000	6000
163	3500	1200	2200	8000
204	4000	1600	2500	10000
255	4000	2000	3000	15000
321	4000	3000	3500	20000

Fig 9.7 Typical general arrangement with dimensions and weights of an oil-injected screw compressor module with horizontal oil separator

to prevent reverse rotation of the compressor due to backflow of gas and is almost always supplied by the vendor. Note also that a pressure relief valve *must* by provided between the isolating valves.

Due to the cooling effect of the injected oil, multi-stage oil-injected compressors do not normally require intercoolers between stages and are sometimes even mounted one above the other with the LP casing discharging directly into the HP casing inlet.

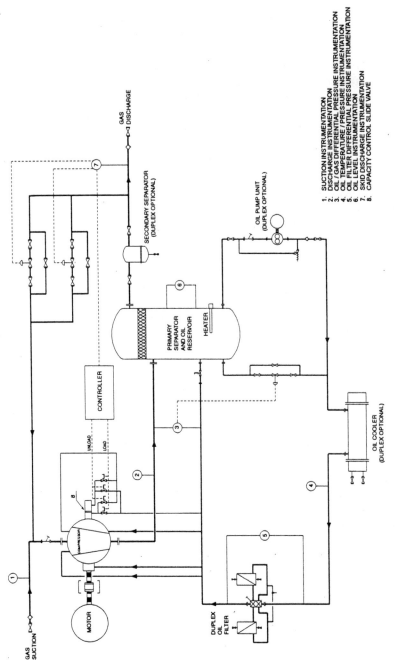

1. SUCTION INSTRUMENTATION
2. DISCHARGE INSTRUMENTATION
3. OIL / GAS DIFFERENTIAL PRESSURE INSTRUMENTATION
4. OIL TEMPERATURE / PRESSURE INSTRUMENTATION
5. OIL FILTER DIFFERENTIAL PRESSURE INSTRUMENTATION
6. OIL LEVEL INSTRUMENTATION
7. SKID DISCHARGE INSTRUMENTATION
8. CAPACITY CONTROL SLIDE VALVE

Fig 9.8 Typical schematic diagram for oil-injected compressor module

9.15 Lubricants

Mineral oils are oxidised by air; the higher the temperature the faster a mineral oil will deteriorate in the presence of air. In bearing systems, only a small fraction of the oil is heated up to bearing temperature at any time and the hottest oil in the bearing has little contact with air. By contrast, in oil-injected screw compressors there is intimate contact between air and the gas that is being compressed. Inhibited mineral oils have a life of less than a year when operating at 85 C with air present. The life drops to less than 1000 hours if the temperature rises to 120 C. Synthetic lubricants can operate at higher temperatures and are being increasingly used. In the absence of oxygen, mineral oils can operate at much higher temperatures though their viscosity drops rapidly with rising temperature. Again synthetic lubricants, some of which have high viscosity indices, can be used over a large temperature range and are favoured for use in screw compressors. Note that any lubricant used in an oil-injected screw compressor must be compatible with the gas handled.

9.16 Oil carryover

At the delivery from the compressor, there is an oil/gas mixture. The oil must be separated, cooled and returned to the reservoir. For many applications, some carryover can be tolerated. For example, it is acceptable and even advantageous, for workshop air, which is used to power pneumatic tools, to contain a small amount of oil carryover. Oil carryover helps to prevent corrosion in the pipework used for distribution. Single stage separators can reduce liquid-phase carryover down to 15 ppm. For better separation, coalescing and occasionally even active carbon filters are needed. If vapour-phase oil carryover is a significant concern (e.g. possible catalyst contamination in the downstream process), it is possible to install an aftercooler between the primary and secondary separator vessels, which has the effect of condensing much of the vaporised lubricant.

9.17 Oil-from-gas separation

9.17.1 Primary separator vessel

The oil recovery/supply vessel, known as the primary separator, is usually a standard dished-end pressure vessel with internal cyclone, baffles and/or wire-mesh demister devices to remove the oil from the gas following compression down to a residual oil carryover of 100–200 ppmv. In process applications, the vessel is usually arranged vertically, is provided with manway access, and is mounted on the compressor unit baseframe. The vessel is normally fabricated in carbon steel, although the demister pack is usually completely in stainless steel. The lower part of the vessel serves as the reservoir for the oil management system and can usually be designed with an oil retention time of

60 to 90 seconds. When particularly corrosive gases are being compressed, it is usual to line the upper part of the vessel (i.e. above the demister pack) with an epoxy or glass-based substance.

9.17.2 *Secondary oil separation*

Where oil carryover in the order of 100–200 ppmv is unacceptable to the downstream process, it is common to install a secondary separator which can reduce the oil carryover (in liquid phase) to below one (1) ppmv. The secondary separator usually employs coalescing filter technology.

Liquid/gas coalescers operate by increasing the size of the liquid droplets as they pass through the porous matrix. The basic principle behind the operation is that of inertial impaction, whereby the inertia of the liquid droplets tends to make them continue on a straight line rather than follow the tortuous gas paths through the coalescer. Consequently, the droplets will make contact with the internal surfaces of the coalescer and run downwards under the force of gravity; as they do so, they encounter other droplets and hence increase in size (i.e. coalesce).

Two steps are involved in correctly designing coalescers; first, the correct number of coalescing elements must be established, based on the flowrate and system physical properties; secondly, the diameter of the vessel must be calculated to ensure that the annular gas velocity is sufficiently low to prevent re-entrainment of the large liquid drops exiting on the exterior of the elements.

Construction of the vessel, as well as lining to combat corrosion, is as for the primary separator. The materials used in the coalescing elements are normally entirely non-ferrous and are chemically inert; these elements have been successfully used with gases containing high concentrations of sulphides and chlorides, as well as with wet carbon dioxide and/or ammonia.

Chapter 10

Twin Shaft, Positive Displacement, Straight-Lobe Blowers

10.1 Introduction

It is assumed that the reader has already been through the preliminary selection process (Part One, Chapter 2) and that a positive displacement (p.d.) blower is the appropriate choice for the duty required. This type of blower was first invented by Jones in 1846. It was reinvented by the Roots brothers in USA and is commonly known as the 'Roots' blower.

10.2 Operation

The p.d. blower consists of two rotors with parallel shafts mounted in bearings with a pair of timing gears at one end to keep the rotors in synchronism. Rotors may have two, three or even four lobes. Normally they have identical profiles but different shapes are possible (see Fig. 10.1).

The rotating shaft traps some gas in the casing as it rotates prior to opening the trapped space to the delivery system. Note that no compression takes place while the gas is trapped but the pressure is increased by gas from the delivery system flowing backwards. The operation is similar to a lock in a canal where the level is raised by allowing water from the upstream system to raise the level (head) in the lock.

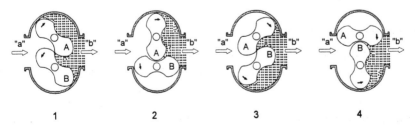

Fig 10.1   Principle of operation of a positive displacement blower

10.3 Application

The main use of p.d. blowers has been to supply low pressure air for, among others, the following purposes: pneumatic conveying of particulate material such as flour and cement, water and sewage treatment and also for providing a moderate vacuum in the paper industry. Recently p.d. blowers have also been used to handle gases and they are also used as gas metering devices connecting a revolution counter to the shaft instead of a motor.

10.4 Range

P.d. blowers are available in a range of sizes from 0.01 to 25 m³/s (30 to 80 000 m³/h). In contrast to other compressors they do not develop a head, nor do they have an in-built volume ratio.

Positive displacement blowers are limited by the differential pressure they can withstand. Only at low suction pressures does the pressure ratio become a limit. The maximum allowable pressure difference depends mainly on the slenderness of the rotor. Any L/D ratio is possible but rotors with L/D values of 1, 1.5 and 2 are the most common ones. They have maximum permissible pressure differences of about 1, 0.8 and 0.5 bar. Note that as the slenderness increases the permissible Δp goes down and the volumetric and adiabatic efficiencies increase due to a relative reduction in the leakage paths.

10.5 Swept volume

The volume of gas displaced by the rotors of a p.d. blower as they mesh is the 'free volume' of the rotors. It is approximately

$$V = \pi/4 D^2 L N \text{ m}^3/\text{s} \tag{10.1}$$

10.6 Efficiencies

10.6.1 Adiabatic efficiency

The adiabatic efficiency of p.d. blowers peaks at a pressure ratio of about 1.2 where it can reach 70 percent. With increasing pressure ratio it will drop to about 50 percent at $\Pi = 2$ (see Fig. 10.2).

10.6.2 Volumetric efficiency

In a p.d. blower all clearances, those between the lobes, the rotors and the end wall and the rotor and the casing, are all subjected to the total pressure difference. The volumetric efficiency therefore drops rapidly from about 100 percent to 70 percent as the pressure ratio is increased from one to two. Further the leakage quantity depends on time, the flow on operating speed. Losses due to bypassing therefore increase with decreasing speed.

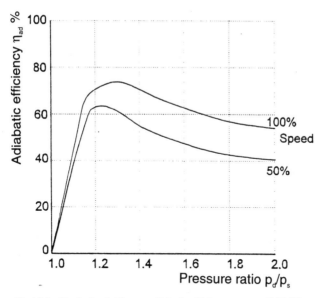

Fig 10.2   Typical p.d. blower adiabatic efficiency curves (O'Neill)

10.7   Tip speed

Typical tip speed for a p.d. blower is between 15 and 30 m/s. Efficiencies improve with increasing tip speed.

10.8   Pressure rating

It has already been stated that differential pressure is the most significant limitation of p.d. blowers. Most applications of these machines are around atmospheric pressure. Such machines have cast iron casings. However, in the process industry there is often a requirement for a circulator operating at raised pressure but requiring only a small pressure rise. Special blowers are available with pressure ratings of 15 and 25 bars with nodular cast iron and cast steel housings respectively. Note that these HP blowers too are restricted to pressure rises of 1 bar.

10.9   Clearances

The p.d. blower relies on small clearances to limit bypassing between its components as there are no contacting seals such as piston rings in reciprocating compressors or vanes in sliding vane compressors. There are no wearing components (if the gas handled is clean). For optimum efficiency internal clearances must be kept as small as possible.

10.10   Discharge temperature

With p.d. blowers the discharge temperature, like pressure ratio, is usually of less significance than the temperature difference between suction and delivery temperature.

Thermal growth can cause reduction of running clearances, leading to interference inside the blower. For surface cooled blowers the maximum permissible delivery temperature is 150 C. Injecting vaporised water into the suction of the machine allows the overall temperature to be kept low. Attention must be paid to the possibility of a corrosive mixture being formed which is unacceptable to the process or necessitating expensive materials of construction.

10.11   Control

The suction flow of p.d. blowers, is constant. The simplest way of control is bypassing though speed variation can also be used. However, even with speed variation there is a significant drop in efficiency at lower speeds.

10.12   Power

The shaft power required for gas compression is

$$P = V_s \Delta p / (\eta_{mech} \eta_{ad}) \qquad (10.2)$$

with V in m^3/s and Δp in pascals. The mechanical efficiency increases from about 94 percent to 98 percent with increasing blower size.

10.13   Scope of supply

The blower manufacturers scope of supply normally includes the complete drive train, including the driver, gearbox, lubrication system and baseframe. Inlet and outlet silencers as well as a suction strainer are usually also included in the supply.

10.14   Materials of construction

The majority of p.d. blowers are low pressure machines used to pressurise air. Cast iron casings and rotors, either forged or cast rotor bodies mounted on forged shafts are suitable materials. For corrosive gases high alloy materials are sometimes required. Special care must be taken when stainless steel is used as it has a tendency to pick up if contact occurs between moving parts.

10.15   Gas contamination

The similarity between p.d. blowers and oil-free screw compressors means that these two types of machine can handle the same levels of contaminants (see paragraph 8.13).

Chapter 11

Rotary, Sliding-Vane Compressors

11.1 Introduction

It is assumed that the reader has already been through the preliminary selection process (Part One, Chapter 2) and that a rotary sliding-vane compressor, hereafter called vane compressor, is the appropriate choice for the duty required.

This type of compressor has been in use for about 100 years, although early examples suffered high internal leakage and wear. Development of the vane design and materials has led to great improvements in reliability and wear patterns. The development of self-lubricating materials made oil-free compression a possibility, but the main development, about 1950 was the introduction of oil-injection into the compression space. Oil-injection allows higher pressure ratios to be achieved and opened up new markets.

11.2 Operation

The compressor comprises a rotor which is located eccentrically in a cylinder of a larger bore. The rotor has several radial slots in which vanes are allowed to slide radially. A typical sectional drawing is shown in Fig. 11.1.

The vane compressor belongs to the family of positive displacement machines; it compresses gas by drawing it in, trapping it and then physically pushing it out of the discharge. Like a screw compressor it compresses gas by reducing the volume, so the inherent property is a volume-ratio, not a pressure ratio. The mode of operation is in three phases, namely: suction, internal compression and discharge.

In position 'A', of Fig. 11.1, the point where the rotor is in closest contact with the cylinder, the equivalent of top dead centre, the delivery period ends. Turning further it will open the space between blades 1 and 2 to the suction line at point 'B'. The space between the blades will increase as they move radially out from the rotor till point 'C', equivalent to bottom dead centre, is reached. A gas volume is not trapped between vanes 4 and 5. As the rotor continues to turn this volume is reduced till at point 'D' the delivery space is opened, and the trapped gas is discharged into the delivery line. At point 'E' the now minute space, the clearance volume, is isolated from the delivery space. An indicator diagram is shown in Fig. 7.8.

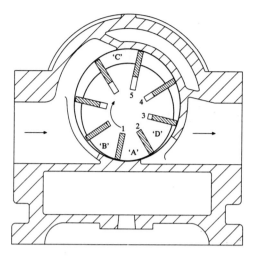

Fig 11.1   Vertical cross-section of vane compressor (A–C Compressor)

The only moving parts in a sliding vane compressor are the rotor and the vanes; no valves are required. Operation is entirely rotary and, although the rotor itself is symmetric about its centre of rotation, the action of the vanes and the pressure involved in forcing the vanes back into the rotor will cause some vibration. Further suction takes place over one sector of the rotor and discharge over another. Due to the change in pressure lateral forces act on the rotor.

The number of vanes can vary considerably. Vacuum pumps may have as few as one or two vanes, whereas high pressure boosters have been built with as many as 24. The vanes must move fully out and in during each rotation of the rotor. Outward movement is by centrifugal force, inward by the force exerted by the cylinder wall. Besides the wear at the tip of the vanes, side forces arise from the gas pressure differential across the vanes as they rotate and lead to wear on the side of the vanes. To reduce the loading on the vanes their number is increased with increasing pressure difference. Further, the smaller the pressure drop across the vane the smaller the leakage across it. The swept volume decreases with an increasing number of vanes. The rigidity of the rotor depends on its 'solid' diameter which must be less then the base circle of the rotor slots.

11.3   Applications

The traditional use of vane compressors has been to supply air for a variety of industrial applications. Nowadays, the higher pressure range of this market, is totally dominated by oil-injected screw compressors and the vane compressor has found new specialised applications within the air compression market as well as many refrigeration and process gas duties. As air compressors they are widely used in waste water and sewage treatment plant, blast hole drilling, plant and process air, particularly in the textile industry. As gas compressors they are used on refrigerants for many industrial refrigeration and vacuum chilling applications, hydrogen vapour recovery, natural gas gathering, fuel gas supply to engines and boilers and to collect landfill and sewage digester gases.

Standard vane compressors are available in a range of sizes from 0.014 to 1.4 m^3/s (50 to 5000 m^3/h) and have a maximum discharge pressure of about 10 barg. They are also ideal for vacuum duties down to an absolute inlet pressure of 0.05 bar.

11.4   Built in volume-ratio

It was noted above that vane compressors have an in-built volume-ratio. Their efficiency is at a maximum when the corresponding pressure ratio results in a delivery pressure equal to that in the delivery system (see Fig. 7.8 and Fig. 11.2).

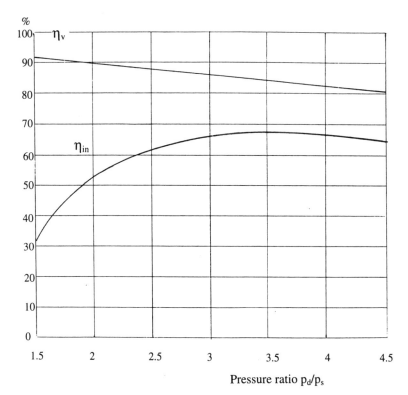

Pressure ratio p_d/p_s

Fig 11.2   Volumetric and internal efficiencies with $\Pi_i = 3.0$, operating at different discharge pressures (p_d)

11.5 Pressure ratio

The internal pressure ratio is dependent on the in-built volume ratio and is:

$$\Pi = (V_i)^\kappa \tag{11.1}$$

Compressors are limited to a compression ratio of about four. Two-stage vacuum pumps can operate from less than 2 mbar to atmospheric pressure giving a stage compression ratio of approaching 30. Such machines usually have water cooled casings.

11.6 Discharge temperature

For uncooled casings the discharge temperature T_d can be estimated from the pressure ratio

$$T_d = T_s \Pi^{(n-1)/n} \tag{11.2}$$

Materials and internal expansion, which leads to reduction in axial clearance, limit the delivery temperature to 140 C, though specials with delivery temperatures up to 175 C have been built.

Vacuum machines usually have water cooled casings to permit the high pressure ratios commonly used.

11.7 Swept volume

The swept volume V_s neglecting the space taken up by the vane is

$$V_s = \pi/4(D^2 - d^2)L \tag{11.3}$$

Allowing for the vanes the swept volume becomes:

$$V_s = \{\pi/4(D^2 - d^2) - zw(D - d)\}L \tag{11.4}$$

Where z the number of blades and w their thickness.

The length of the rotor, in relation to its diameter varies between one and two. Compressors with short rotors can withstand larger pressure differences, those with long ones are more efficient as the end losses are lower. They are also cheaper for a given capacity.

11.8 Discharge pressure rating

Although a vane compressor is a low differential pressure machine, its discharge pressure capacity is determined by its casing design and material. Since most vane compressors operate at or about atmospheric pressure, the casing need only be capable of withstanding vacuums or low positive pressures, and is usually in grey cast iron. In the process industry, however, there is often a requirement for a low Δp machine operating at elevated pressures as a circulator. For these applications specials with nodular cast iron casings are available from some manufacturers.

11.9   Rotor tip speed

Vaned compressors are slow running machines usually directly driven. Average tip speed is usually between 20 and 30 m/s. Higher speeds result in rapid wear of the vanes, lower speeds decrease the efficiency due to increased bypassing.

11.10   Vane compressor package

Vane compressor manufacturer's scope of supply normally includes the complete drive train, including electric motor, gearbox (where required) and couplings. An oil supply console, which may be integral with the compressor base frame or a separate unit, to supply lubricating and seal oil, as well as instrumentation and auxiliary equipment, form part of the complete package.

The vane compressor module has to be fully integrated into the process system. Care must be taken to ensure that the compressor receives gas at the design conditions. Particulate contamination will result in wear and erosion, and it may be necessary to install a knock-out drum upstream of the machine.

11.11   Materials of construction

Major items are supplied in the following materials:

(a) Casings

- grey cast iron for air and most gases
- cast SG (nodular) iron for hydrocarbon and other hazardous gases where site conditions/specifications will permit a less expensive material than steel
- cast carbon steel for other applications (although this is rarely seen in practice).

(b) Rotors

Note that vane compressor rotors and shafts are usually machined from bar stock for smaller sizes and integral forgings for larger sizes.

- carbon steel for air and most gas applications
- high-alloy steel (typically 12–13 percent Cr, 4–5 percent Ni) for corrosive gases, sour gases (to meet NACE and API hardness and yield strength requirements) and wet gases or gases requiring water-injection.

(c) *Vanes*

– Typical for process gas service is laminated cloth, impregnated with phenolic resin and heat treated.

11.12 Gas contamination

11.12.1 *Particulate matter*

Some gases, e.g. coke-oven gas, contain hard particulate matter entrained in the gas stream. In such applications, to resist abrasive erosion, the gas should not be permitted to contain hard crystalline particles larger than $50\,\mu$m. In most cases the gas is at least partly cleaned at source, although it may be necessary to install a cyclone-type separator upstream of the compressor inlet.

Other gases, e.g. lime-kiln gas in a soda ash plant, contain soft particulate matter, entrained in the gas; unless corrosive, this type of particulate presents few problems for a vane compressor. The continuous 'sweeping' of the vanes actually prevents heavy build-up and ensures even distribution of deposited material. Most process gases are sufficiently clean, although in some cases a condensing cooler and an effective droplet separator should be provided upstream of the compressor inlet.

Nominally clean, dry gases also need careful evaluation: permanent inlet gas filters may be needed where products of corrosion of preceding equipment can be carried forward or where catalyst dust may be present.

11.12.2 *Droplets in suspension*

Liquid droplets present in the incoming gas, will cause erosion; to avoid erosion, a droplet separator should be provided, upstream of the blower, with an effective cut-off size rating of 25 μm.

Where liquid is deliberately injected into the gas stream for cooling or washing purposes, it is important to atomise the liquid at entry in order to avoid similar erosion problems.

11.12.3 *Molar mass variation*

Wide variations in molar mass have little effect on vane compressors, although significant variations in the isentropic exponent (κ) of the gas can lead to unacceptably high discharge temperatures.

11.12.4 *Compressibility variation*

Operation close to the saturation line presents few problems for vane compressors, and even two-phase flow is acceptable in this type of machine (liquid-phase content of up to 10 percent by volume is handled without difficulty). Care must be taken, however, to avoid large liquid droplets (or 'slugs' of liquid) entering the compression space and these should be dealt with as in paragraph 11.12.2.

11.12.5   Oxygen-rich gases

Pure oxygen or gas mixtures containing more than 50 percent oxygen are not suitable for compression in vane compressors since these are normally oil-lubricated.

11.12.6   Gas solution in lubricating oil

Vane compressors normally use a 'once-through' oil injection system, with the oil discharging together with the gas. Oil separation systems, as supplied for oil-lubricated reciprocating compressors, can be provided if required. Oil with gas dissolved in it is not reused as a lubricant.

Part Four

Common Features

Chapter 12

Drivers and Transmissions

12.1 Introduction

The driver selection for any fan or compressor is based on factors such as utility availability and need for energy conservation. Each driver type will have specific design and operating limitations and it is important that these limitations are identified during the early design stage. This chapter describes the applications and limitations for drivers which are used to drive compressors and fans.

As a guide this chapter also covers the limitations of key components in the drive train such as the gearbox and shaft couplings.

12.2 Factors influencing driver selection

The key factors influencing the choice of driver type are usually the utility availability, process operation and control requirements, and the integration of energy systems on the plant.

Process operation requirements are often overlooked, but may have a critical influence on driver selection. For example when driving a centrifugal compressor, molecular weight variation of the process gas may require reduced speed operation because of a discharge temperature limitation in the compressor. Also during startup, the available driver starting torque should exceed the torque requirements of the compressor for all process gases specified.

The influence of utility availability is of prime importance. The allowable voltage drop in an electrical distribution system, for example, may limit the motor size or require a reduced voltage start system.

12.3 Motors

The choice of motor type will affect the selection of the driven machine. The use of a variable speed motor, on a voltage limited electrical distribution system, will limit the voltage dip during motor start. On a sensitive distribution system this can be an important consideration. Similarly, the gear ratio for a fixed speed motor and driver may require a gearbox which is not commercially

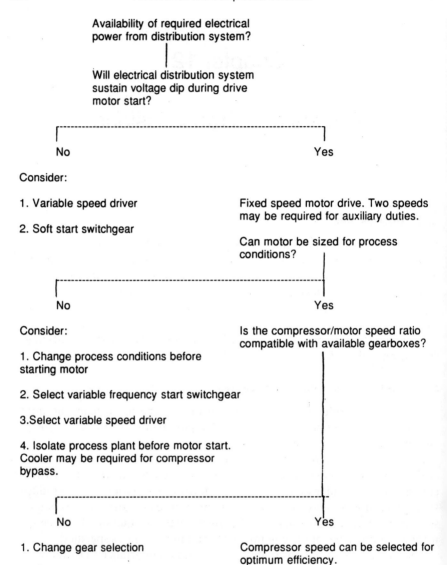

Availability of required electrical
power from distribution system?

Will electrical distribution system
sustain voltage dip during drive
motor start?

No   Yes

Consider:

1. Variable speed driver   Fixed speed motor drive. Two speeds
                                            may be required for auxiliary duties.
2. Soft start switchgear
                                            Can motor be sized for process
                                            conditions?

No   Yes

Consider: Is the compressor/motor speed ratio
                                            compatible with available gearboxes?
1. Change process conditions before
starting motor

2. Select variable frequency start switchgear

3.Select variable speed driver

4. Isolate process plant before motor start.
Cooler may be required for compressor
bypass.

No   Yes

1. Change gear selection   Compressor speed can be selected for
                                            optimum efficiency.
2. Change compressor speed - this may
mean reduced efficiency operation

Fig 12.1   Typical decision tree for compressor motor selection

available or cost effective. The driver speed, and hence driver selection, may
have to be reconsidered.

Figure 12.1 is an example of a typical decision-making process used to select
a fixed speed gas compressor motor driver and compressor arrangement.

12.3.1   *Fixed speed motors*

Motors may be of the induction or synchronous type as described below.

Induction motors
Induction motors operate at a speed slightly below the synchronous speed. The difference, known as slip, is small—in the order of 1 to 1.5 percent and is directly proportional to the load applied. As a result induction motors will lose speed if large loads are applied. Startup can be directly on line, or to reduce startup current, via some startup device like a star/delta starter, or via a transformer.

Synchronous motors
A synchronous motor, as its name implies, operates at constant speed. Two characteristics of synchronous motors require special attention:

(a) *Startup.* To bring these motors up to speed either a pony motor or other driver is required, or if they are to be started direct on line, then they must operate as induction motors during the speed up period. This mode of operation, particularly with salient pole motors, causes large torque fluctuations with a frequency twice the slip frequency. A torsional resonance will occur which can lead to torques many times larger than the full load torque. To avoid the possible effects of this resonance a dampening element, such as a rubber coupling, is required.

(b) *Excitation.* The rotor of a synchronous motor is a large electromagnet and needs to be supplied with a large DC current—in the order of 1000s of amps.
There are two common methods:

 – slip rings with carbon brushes. These spark in operation and for safety reasons they may require purging. Also the carbon brushes wear and need to be renewed every 4000–5000 hours, which requires the machine to be shutdown
 – brushless excitation.

Motor speed is fixed, by the frequency of the electrical system and the number of motor poles.

There are other design considerations which may affect the choice of fixed speed motor:

 – for two pole motors in excess of 3 MW power rating the rotor lateral analysis must be studied carefully to avoid resonance and motor vibration
 – a gearbox is invariably required when the speeds cannot be directly matched. This may add a substantial lubrication system, additional maintenance needs and extra layout space. Damage can occur to the gearbox if the system produces negative torque fluctuations
 – a vee belt drive is generally not used with a synchronous motor

- for motors in excess of 1 MW allow access for rotor removal in situ, together with an adequately sized crane.

12.3.2 *Variable speed motors*

There are many ways of varying the speed of an electric motor. Some common methods are:

- DC motors which require commutators and brushes
- induction motors with wound rotors. Only speeds below synchronous are possible. They are particularly useful for applications where the required torque drops rapidly with speed with, for example, roto-dynamic machines
- variable frequency drives. The available torque is constant over the whole speed range, although special cooling must be provided if extended full torque operation at low speed is required.

There are other design considerations which may affect the choice of variable speed driver:

(a) Variable frequency drives give excellent speed control. This may simplify the process control system, and allow improved operating efficiency.

(b) Variable frequency units are dimensionally large, will generate heat (heat load is about 3 percent of motor rating) and adequate layout space for the substation and ventilation must be provided.

(c) Co-ordination between the suppliers of the motor and the variable frequency unit is critical. The wave form generated by the variable frequency unit may be non-sinusoidal and may cause overheating of the motor rotor. The motor is normally derated to take this into account, or alternatively more reactance must be added to the system.

(d) Overspeed speed protection is necessary to protect against the risk of uncontrolled overspeed. This is normally provided by a frequency limiter in the variable frequency unit.

Figure 12.2 illustrates approximate speed and power limitations on the application of variable speed and variable frequency motors.

12.3.3 *Starting torque*

The driver power characteristics are critical to the design of the process system. The required driver power characteristic will be influenced by the choice of the driven machine, the control system, the process system isolations (see Fig. 12.3) and the operating limitations of the driven machine.

For a fixed speed motor, the starting torque available will vary according to the motor design as well as the voltage available from the electrical distribution system which will fall due to the large current taken during startup.

A variable speed driver will provide constant starting torque, this may be

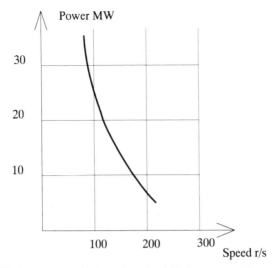

Fig 12.2   Current limitations on variable speed and variable frequency motor applications

advantageous where the driven machine starting torque requirements are arduous or where the electrical distribution system is sensitive to load.

Figure 12.3 shows two typical induction motor torque/speed curves, one for full voltage at the motor terminals and one assuming a small voltage drop. Curves are also shown for a number of starting configurations of a gas compressor.

– Curve 1 represents the required starting torque with a suction throttle valve on the gas compressor to reduce the suction pressure, with the compressor isolated from the delivery system.
– Curve 2 represents the required starting torque with inlet guide vanes closed to minimum position, with the compressor isolated from the delivery system.
– Curve 3 represents the required starting torque without suction throttle or inlet guide vane, with compressor isolated from delivery system.
– Curve 4 represents the required starting torque without suction throttle or inlet guide vane with compressor at full discharge pressure.

In the case illustrated the available torque from the motor is only adequate in Curves 1 and 2. The compressor must either be isolated from the discharge system, hence influencing the design of the process isolations, or the pressure of the discharge system must be reduced before starting the compressor.

12.3.4   *Motor speed*

Where a speed increasing gearbox is required between a fixed speed motor driver and the driven machine, a four pole motor is generally preferred. This is

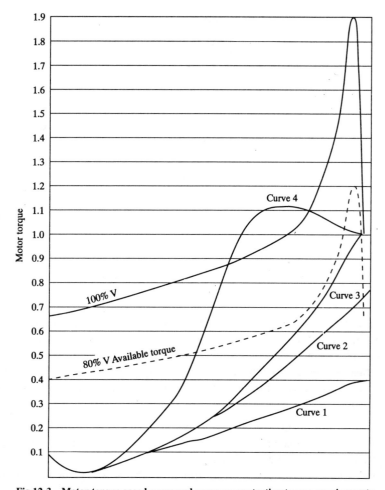

Fig 12.3 Motor torque speed curve and compressor starting torque requirements

because four pole shaft stiffness will be greater, reducing the possibility of a lateral resonance in the motor.

12.4 Steam turbines

Process waste heat is usually used to raise steam and if sufficient steam is available, a steam turbine may be selected. A single large turbine may be preferred to drive an alternator or alternatively a process compressor. Turbine drives are ideally suited for speed control and also give full torque at any speed. The most efficient turbines for a given duty and steam supply usually do not match the speed required by the most efficient compressor, and a gearbox may be required.

Some common configurations of steam turbines used in the process industry are:

- Back pressure
- Condensing
- Extraction
- Pass-in/pass-out

Described below are some common steam turbine features and limitations. Specific limitations for each turbine type are outside the scope of this book and should be discussed with the manufacturer.

12.4.1 *Blade type*

Two blade types are commonly found: impulse or reaction blades. The impulse blade design is generally considered to have a more robust construction, but may be three to four points less efficient than the reaction machine. Roto-dynamically the impulse blade type is usually a flexible shaft design, and therefore speed matching the turbine with the driven machine is likely to be limited by the forbidden speed bands in the turbine.

12.4.2 *Starting torque*

The steam turbine offers advantages on startup due to its constant torque characteristic which is independent of speed.

12.4.3 *Speed matching*

Ensure that the turbine selection is based on proven mechanical drive experience if the use is not for fixed speed power generation. Blade stress levels need to be evaluated and limited to known previous experience, particularly in the wet/dry transition zone.

One factor in sizing the turbine will be the limitation on blade tip speeds. Multiple exhaust design can overcome this problem, allowing the limiting exhaust end blading to be reduced in diameter, but maximising the exhaust steam volume.

12.4.4 *Inlet steam conditions*

Check supplier experience for extremes of inlet steam temperature and pressure, back pressure and extraction rates.

Check the steam purity values used in the design basis for blade stresses calculated by the manufacturer. Any deviation in steam purity from these values will require a review of blade stresses, blade profile and number of

Table 12.1 Maximum steam impurities at turbine inlet for
steam pressures >64 bar

Conductivity	μS/cm	<0.20
Silica (SiO_3)	mg/kg	<0.02
Iron (Fe)	mg/kg	<0.02
Copper (Cu)	mg/kg	<0.003
Sodium (Na)	mg/kg	<0.01

stages within the turbine. For guidance a German requirement for purity of high pressure steam (>64 bar) is shown in Table 12.1 (reference (21)).

12.4.5 *Extraction*

Ensure the steam flow to the condensing section on an extraction/condensing turbine is adequate to remove the heat due to windage.

12.4.6 *Blade resonance*

Speed variation will generally be limited by blade resonance. Ensure that natural frequency modes for roto-dynamic torsional and lateral critical frequencies for the machine system and the blades do not coincide. Torsional frequencies of the machine system are usually much lower than the blade frequencies.

Turbine blades need to be designed to avoid the possibility of resonance and early failure and the predicted response of blades to induced vibration is normally presented graphically. Campbell diagrams show blade and excitation frequencies. These indicate the margin between the blade harmonic and any exciting forces. To avoid resonance, coincidence between these two must be avoided. Campbell diagrams show the relation between tolerable steady and fatigue load of materials to avoid failure.

A number of techniques are available to the blade designer to prevent unacceptable dynamic blade stresses. Increasing damping by using blade shrouds or lacing wires or using increased blade profiles are a few remedies.

12.5 Gas turbines

12.5.1 *Types*

Industrial/Aero derivative types
Aero derivative gas turbines are being increasingly used in industrial applications. They are not as robust as industrial machines, and major maintenance is usually by replacement of the engine. Both industrial and aero derivative machines have similar efficiencies.

Single shaft/multi shaft
The advantage of the single shaft design is its simplicity, as the gas turbine

compressor is driven at the same speed as the output shaft. This is normally only economic with constant speed drive machines, e.g. an alternator drive. For other duties it is usual to use a two shaft design such that the compressor speed is independent of the gas generator output shaft.

12.5.2 Starting torque

Like the steam turbine, the two shaft gas turbine has a constant torque characteristic which is independent of speed.

12.5.3 Fuel

Liquid or gas fuel can be used as a source of input heat, however, many turbines are designed for only one type of fuel or a particular fuel composition. The fuel supply may need to be treated to remove substances that could cause corrosion. The fuel supply must also be injected at the correct pressure to suit the gas turbine.

12.5.4 Design conditions

Gas turbines are designed to an ISO rating which gives the power output at the design inlet air condition. Variations in inlet temperature and pressure will change the power output. The manufacturers quoted output should allow for any inlet or outlet pressure drop, as such a pressure drop will reduce the power output.

Gas turbine efficiency is significantly affected by air quality, and inlet air filtration is essential to achieve continuing efficient operation. Specific advice may be required to match the ambient operating conditions with the requirements of the gas turbine.

Exhaust systems are likely to need a silencer, and this must be taken into account in the pressure drop evaluation.

12.5.5 Heat recovery integration

Some of the input heat used producing shaft output power can be recovered by use of a waste heat recovery boiler. Exhaust gas pressure drop across the waste heat recovery system reduces shaft output power. Exhaust gas dampers or diverters may be required to minimise the effect of startup on the heat recovery system. They need careful selection as they operate in a hostile environment.

12.5.6 Exhaust emission

Exhaust emission standards are usually controlled by local or national authorities. NO_x emission levels can be reduced by fitting special combustors or by steam injection for which high purity water is usually required to avoid turbine corrosion.

12.6   Engines

Like reciprocating compressors, a reciprocating engine will require routine maintenance. This may necessitate an installed spare machine.

There are design considerations which may affect the choice of engine. Some key considerations are:

(a) The suitability of a fuel oil-fired engine should be assessed for hazards in the area where the engine is to operate.
(b) Fuel characteristics, gas or liquid, must be assessed to ensure supplies to the engine at the required conditions. A fuel supply system may be needed to supply diesel fuel for engine start and shutdown, and have fuel oil when running.
(c) Operating speeds typically cover the range 10–20 r/s (500–1000 rpm). Higher operating speeds will result in higher maintenance costs, but require a smaller size machine or fewer machines.
(d) Include the costs of air intake and exhaust equipment that will be required.
(e) Care must be taken to match the torque/speed characteristics of the engine to the driven machine.
(f) Reciprocating forces will create torsional oscillations which need to be assessed to avoid resonance with other frequencies. A torsionally flexible coupling may be required and a detailed torsional analysis carried out to clearly define the speed ranges where operation is possible.

12.7   Gas expanders

12.7.1   Design throughput

A key issue in the selection of a gas expander is the design throughput and control method selected to achieve the required process duty. For example, to be able to obtain maximum power from the process, the inlet of the expander should not be restricted under normal operating conditions. Part load regulation may be by partial admission if the gas is sufficiently clean, or by throttling the inlet. An estimate will need to be made of the operating time spent at the various duty points to establish an appropriate regulation method to optimise machine cost and energy conservation.

12.7.2   Overspeed protection

The expander will require overspeed protection. A speed sensing device normally actuates an emergency stop valve on the expander inlet—however, this is dependent upon the ability of the valve to close quickly enough. The overspeed sensing device should be an electronic voting system capable of on-line verification.

12.8 Driver specification

Once the driver and driven machine types have been selected, it is important to ensure that the two are compatible. This section covers the interface between the driver and the driven machine.

12.8.1 Driven machine requirements

Check that the following driven machine requirements are specified:

- power absorbed at normal and maximum conditions
- speed, normal and maximum
- direction of rotation
- startup time
- startup torque
- inertia of driven machine
- area classification

12.8.2 Driver rating

For fixed speed motor drivers the motor starting torque curve will be required to ensure that the startup torque of the driven machine is exceeded during run up of the train.

The driver power rating is usually based on the maximum flowsheet duty. Startup demands may exceed this and a margin may be required according to the type of driver selected. A margin of 10 percent over the driven machine requirements is normally used for motor drivers. Gas turbine drivers need careful selection depending upon the design atmospheric conditions. Do not design for the most adverse conditions, but allow for some reduction in output under these conditions to avoid excessive driver sizing.

Variable speed motors are in principle capable of producing maximum torque at any speed. However, the cooling effect from shaft mounted fans will drop rapidly with speed and therefore full torque (continuous) operation at lower speeds may only be for limited periods. Motors cooled by external fans can operate continuously at low speed and should be specified when a large turndown is required.

12.9 Gearbox selection

A limitation on the selection of the driven machine is the availability of a suitable gearbox which can provide the required gear ratio between the driver and driven equipment.

Some basic limitations on commercially available gearboxes are specified below. Special gearboxes may exceed these limits, and will need careful evaluation before selection.

Determination of gear size

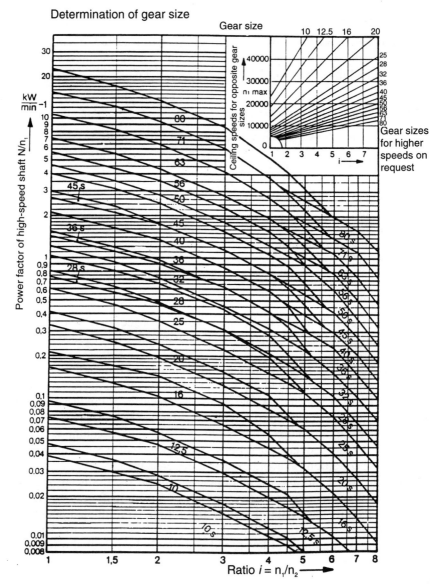

Fig 12.4   Typical gear selection chart: Parallel shaft gears pinion and gear hardened service factor = 1.0

12.9.1   Parallel shaft gears

See Figure 12.4 for a typical selection chart for hardened gears. Note that a factor needs to be applied depending on type of service. Refer to AGMA standard (22) for advice.

12.9.2   *Epicyclic gears*

These gears are not commonly used on fan or compressor applications. Where an application is considered it is better to consult a specialist manufacturer.

12.9.3   *Integral gear*

Compressors with integral gears are normally driven by a large diameter bull gear matched to pinion wheels each of which run at different speeds. The speed of each pinion shaft, and direct coupled compressor wheel, is optimised for maximum efficiency.

12.10   Shaft coupling characteristics

Couplings between the driver and driven shafts may have more than one function, where for example, the coupling also acts as a variable speed device. The coupling selection may have implications on other auxiliaries in the machine train—an oil lubricated gear coupling will require special oil filtration.

Shaft couplings therefore need to be selected for specific characteristics required by the driven machine or the machine system. Typical characteristics are:

– shaft misalignment capability
– soft start or torque limitation
– cyclic torque fluctuation
– axial movement capability/limitation
– axial thrust load

There are four major types of couplings:

(a) *Couplings with rubber elements*. These are suitable for low speed duties. They do allow a small amount of shaft misalignment but their main advantage is that they provide damping of any torsional mode.
(b) *Gear couplings*. These allow some misalignment, indeed they need some misalignment to allow lubrication. They are best used in pairs with a spool piece between the two couplings. They can be greased or oil-lubricated.
(c) *Diaphragm couplings*. Alignment is critical on these, and they need no lubrication.
(d) *Rigid couplings*. Also require no lubrication, and all axial loads are transferred to one bearing, usually on the driver.

Selection of specific coupling types is beyond the scope of this guide, but is covered in an accompanying IMechE guide (**23**).

Chapter 13

Lubrication

13.1 Introduction

All the machines dealt with in this guide are depending on lubricated bearings. Only the smallest fans and compressors rely on grease lubricated rolling element bearings. Some small machines have self-contained oil-lubricated bearings. However, nearly all machines are oil-lubricated, the oil being supplied from a dedicated oil console. This console is usually incorporated in the supply of the compressor or driver vendor. A large percentage of non-availability of compressors can be attributed to failings in the lubrication system. For this reason special care should be devoted to their design and operability.

13.2 Duty

Oil not only provides a film that keeps moving parts apart and so prevents wear, but also removes the heat generated in bearings. Lubricating oil is also used to act as a sealing media in many types of seal and is frequently used as a hydraulic fluid to operate controls.

13.3 Oil unit

13.3.1 *Oil tank*

With small machines the oil tank is frequently used as a base for the whole machine (see Fig. 8.5). Large machines have free standing oil tanks. API 614 (**19**) requires that the tank should have a retention time of 8 minutes, to allow the return oil to be degassed. With high powered compressors this may require oil tanks with a capacity of 10 m^3 or even more. Oil has to be supplied to bearings at about 40 C. To allow the oil pumps to work and machines to be started oil needs to be warmed up. Oil heaters are therefore required. These are often installed in the oil tank. Note that the tank needs sufficient free capacity to be able to accommodate all the oil from all overhead tanks as well as the pipelines when the oil pumps are stopped.

Table 13.1A

Specification for lubricating oil for ammonia synthesis plant compressors

Required characteristics		Test method
Viscosity, kinematic cs at 40 C	ISO VG* as specified by machine supplier	IP71/ASTM D445
Flashpoint, Pensky Martens, closed	min 168 C	IP34/ASTM D93
Pour point	max −6 C	IP15/ASTM D97
Demulsification number	max 300 sec (ISO VG 32*) 360 (ISO VG 68*)	IP19
Rust-preventing characteristics		IP135/ASTM D665
Procedure A	pass	
Procedure B	pass	
Foaming tendency		IP146/ASTM D892
Foam ml at 24 C	max 600	
at 93.5 C	max 100	
at 24 C after 93.5 C test	max 600	
Foam stability after 10 minutes		IP146/ASTM D892
at 24 C	nil	
at 93.5 C	nil	
at 24 C after 93.5 C test	nil	
Air release value		IP313
Minutes to 0.2 percent air content at 50 C max	5 (ISO VG 32*) 11 (ISO VG 68*)	
Oxidation characteristics		IP157/ASTM D943
Minimum time to attain acidity of 2 mg KOH/g	10 000 hours	
Inertness to ammonia		Method A, below
Acid number after test mg KOH/g max 0.1	not detectable	
organic sludge (heptane insolubles)		

*ISO 3448—1975 (E) Industrial liquid lubricants—ISO viscosity classification

Method A Inertness to ammonia

Test method ASTM D943 modified as follows:

a use mixture of air (8.6 litre/hr) and ammonia (1.4 litre/hr) in place of oxygen.
b use copper catalyst only.
c add 1 ml of water to oil (300 ml) at start of test and subsequently at 24 hour intervals.
d maintain test cell at 125 C.
e use test duration of 150 hours.

Notes

1 The sample should be stripped of ammonia by purging with nitrogen before determining the acid number.

2 Organic sludge is specified to differentiate from any copper reaction products formed with the catalyst. No quantitative limit for the organic sludge has been determined, but no problem has been found in clearly differentiating between oils with additives that react with ammonia and those that do not.

Table 13.1B

Routine monitoring of ammonia and methanol plant compressor oil

The following test routine is recommended:

1 Visual examination   Weekly

The sample should be taken in a clean glass bottle and compared with a reference sample (a) and the sample taken the previous week. The points to look for are:

Observation	Action
Change in colour	Investigate (see below)
Free water (cloudiness)	Check centrifuge or coalescer filter (b).

Notes

a The reference sample should be stored in the dark; otherwise it will gradually darken and eventually deposit sludge.

b Combined compressor/steam turbine lubrication systems should be equipped with a permanently installed centrifuge or coalescer to remove free water.

2 Viscosity at 40 C (Test method IP71/ASTM D445)   Three monthly or on

Change oil or discuss with oil supplier if viscosity alters by more than ±10% from new oil value. result of visual examination.

3 Acid number (Test method IP1)   ,,

Change oil if acid number exceeds 1.0 mg KOH/g.

4 Water content (Test method ASTM D1744)   ,,

Steps should be taken to maintain the total water below 200 ppm.

5 Anti-oxidant content   ,,

There is no standard test for the determination of anti-oxidant level. The method given in Analytical Chemistry 1964, Vol 36, p 185 has been found to be suitable for those oils containing 2.6 Di-tert-butyl-p-cresol (dbpc), which applies to most of the oils offered for use in ammonia plant compressors. The level of dbpc in new oils normally ranges from 2500–8000 ppm. The level should be maintained above 500 ppm by addition, as necessary, of a suitable additive concentrate obtained from the oil supplier.

6 Rust-preventing characteristics (Test method IP135/ASTM D665). ,,

The oil should pass Procedure A of the test. If it fails, it should be re-fortified by addition of a suitable additive concentrate obtained from the oil supplier.

13.3.2 *Pumps*

Positive displacement pumps are most frequently used. Lubricating oil needs to be supplied at a pressure of about 2 bar. Seal oil may be required at pressures up to 100 bar or even higher. Seal oil pumps usually take their suction after the lubricating oil coolers and filters. On single stream units always and on other large units the pumps are duplicated, one running one spare. Oil pumps driven off the compressor shaft are common on some types of machine. They have the advantage of supplying oil as long as the main machine is turning and no special run down provisions need to be made.

13.3.3 *Filters*

On single stream machines duplex filters are always used, designed so that they can be switched over without interrupting the oil supply to the main machine. In the past filtration to 40 μ was considered to be sufficient. Nowadays it has been realised that the finer the filtration the more trouble free the system will be. Filtration to 10 μ is common and even 3 μ is sometimes specified. Cleaning filter elements is difficult and throw away cartridges are standard except in locations where such cartridges are not available.

13.3.4 *Coolers*

On single stream machines duplex coolers are always used. Like filters they have to be designed so that they can be switched over without interrupting the oil supply. Coolers should be designed to TEMA (**24**) and have an adequate fouling factor. Particularly in cold climates it is an advantage if the cooler can be operated with hot water to allow rapid heating of the lubricating oil.

13.3.5 *Run down*

In case of power failure lubricating oil must be supplied to all bearings to allow the machine to come to rest. Further, on hot machines, it is essential to keep a small oil flow till the hot shaft is cooled down to about 110 C to avoid damaging the bearing lining. The flow required for these two duties is considerably less than the normal oil flow. A small steam or battery driven pump is required for cooling duty. For rundown overhead tanks are the most economic solution.

Seal oil must be supplied till the pressure inside the casing is released. API 614 (**19**) requires overhead tanks with a minimum capacity of 10 minutes normal flow.

13.4 Oil

It is essential that the oil used is suitable for all the equipment, compressor, driver, gearbox, couplings etc., supplied from the oil unit. Usually a mineral or synthetic oil with a viscosity of 32 or 46 cS to BSI 4231 (**25**) is most suitable.

It is important to agree an oil specification with the vendor of the machine based on a standard rather than select a proprietary brand. In Tables 13.1 (a) and (b) the requirements of one user for oil to be used on ammonia plants is given.

To ensure reliable service the oil should be kept free of water, particularly difficult if the oil is also used to lubricate a steam turbine driver. Centrifuging or vacuum applied to a sidestream are most effective ways to keep oil dry. Periodic tests to confirm that the oil is still suitable for service should be carried out. Such tests are also most useful in detecting machine problems, particularly wear in any of the moving parts.

Oil additives can cause problems, particularly if the oil comes in contact with process gases. Additives should be kept to a minimum and previous experience should be used in selecting a suitable oil for process gas compressors.

Oil is also a nutrient for yeast and bacteria. To prevent their growth in oil systems the oil should be kept water free and warm in the oil tank (>40 C).

Chapter 14

Seals For Rotating Machines

14.1 Introduction

The purpose of a compressor is to raise the pressure of the gas handled. This implies that there are different pressures within a machine. The purpose of seals is to minimise, or even eliminate, leakage flow from high pressure zones to low pressure ones. To achieve this the clearances between rotating and stationary components are kept to a minimum. However, if for some reasons such components with cylindrical shape touch a large amount of heat will be generated due to friction. This heat will tend to bow the rotating member and so cause the rub to increase and consequently cause total failure.

14.2 Labyrinth seals

To reduce the heat developed by friction the minimum clearance is applied only in selected positions by using a labyrinth (see Fig. 14.1). This has a double advantage:

(a) Firstly, the flow through a multi-stage labyrinth is less than that by a single narrow gap and the reduction in flow is approximately inversely proportional to the square root of the number of seal strips. The gas passes through the narrow gap into a larger chamber and so loses its kinetic energy before passing through the next restriction. A formula giving the flow through a multi-stage labyrinth is:

$$m = \varepsilon \zeta A p_s / \sqrt{(10^3 R T_s)} \tag{14.1}$$

where A = free area in m^2 between stationary and moving parts.

Fig 14.1   Labyrinth

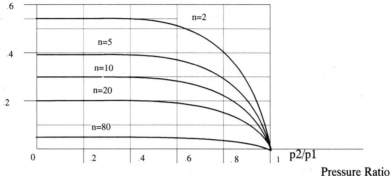

Fig 14.2   Labyrinth leakage factor ε

Pressure Ratio

Figure 14.2 shows ε, a factor depending on the pressure ratio r and the number of tips n. It is usable with critical as well as sub critical flow. ζ is a factor depending on the shape of the labyrinth. It varies between 0.6 and 1.5. The lower values apply to well staggered sharp seal strips.

Figure 14.2 also shows that an increasing number of labyrinth tips has a diminishing return but this is only partly true. Wear is always greatest in the last stage where the gas is at its lowest pressure and hence the largest velocities occur, particularly with sonic flow. Once the last strip is worn (i.e. the clearance has increased) the sealing function is transferred to the preceding labyrinth tip so that an increased number of labyrinth tips ensures a longer life with small leakage.

(b) Secondly, the heat generated in case of a rub is minimised, particularly if the labyrinth is made of soft metal (e.g. silver or aluminium), though a rub with steel labyrinths on a rotor usually leads to a rotor bend and serious damage.

Labyrinths are the standard method of preventing recirculation inside multi-stage compressors. They are also used to seal compressor shafts entering into casings if some leakage can be tolerated e.g. on air compressors. A method frequently used with gases other than air is to remove any leakage with an ejector. Ejectors are placed such that they suck any gas leaked from a chamber close to the atmospheric end of the labyrinth. Some air is also sucked in so ensuring that there is no outward gas flow at the open end of the labyrinth.

14.2.1 *Balance piston*

In centrifugal and axial compressors the difference in pressure acting on the inlet and exit side of each stage causes a large axial load, particularly with high pressure compressors. This load is relieved by the use of a 'balance piston' (see Fig. 4.8). This device is a sturdy labyrinth, often with a honeycomb rather than a strip labyrinth. The final delivery pressure is on one side and the suction pressure on the other side of a drum with a diameter such that the balance force is of the same order as the sum of the unbalance forces of the individual stages but in the opposite direction. On high pressure machines the unbalance forces have to be lowered by a factor of ten, or even more, so that the thrust bearing can carry the axial load.

14.3   Positive seals

To provide a 'positive' seal three methods are in use. It should be noted that though these methods are known as 'positive' a small amount of leakage always does occur and is usually dealt with in an auxiliary system.

14.3.1   Mechanical seal working on barrier liquid

A mechanical seal has a moving and a stationary face at right angles to the shaft axis (see Fig. 14.3). The faces are extremely flat (within two or three light bands) and though spring and pressure loaded, are kept apart by a fluid film in operation. There is no contact between the faces when operating hence there is no wear and the life of mechanical seals is very long.

Double mechanical seals with a barrier fluid, usually lubricating oil, between

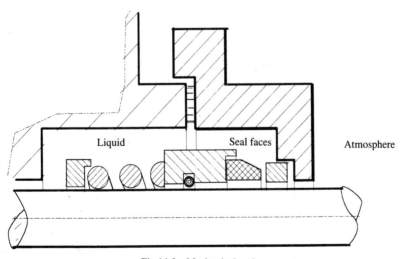

Liquid   Seal faces   Atmosphere

Fig 14.3   Mechanical seal

them is a frequently used method to prevent gas leakage at pressures up to about 30 bar, though sometimes they are used at higher pressures. A small amount of barrier fluid (cc's /hour) passes the inner seal and has to be separated and either returned to a reservoir or in a case of serious contamination, e.g. by sulphides, discarded. Mechanical seals are particularly popular on refrigeration compressors where it is customary to have all bearings in the refrigerant and only one mechanical seal is needed where the drive shaft enters the compressor. Note that mechanical seals provide a seal when the machine is at standstill even if there is no supply of barrier fluid. Additional mechanical seals are sometimes installed as standstill seals. The faces of standstill seals are normally kept apart and close only in the absence of oil pressure.

For details of mechanical seals see (**26**).

14.3.2 *Floating rings*

Floating rings (see Fig. 14.4) like double mechanical seals, work with a barrier liquid, usually oil. This liquid is injected between two floating rings at a pressure marginally higher than the gas pressure on the compressor side. To keep the oil flow to a minimum the clearance between the rotating shaft and the ring is smaller than the bearing clearance. It is for this reason that the rings are floating. If they were fixed they would act as bearings and rapidly fail. Note that the rings, while free to float, are prevented from rotating. Floating ring seals

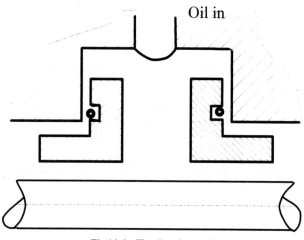

Fig 14.4   Floating ring seal

can be used up to very high pressures, 400 bar and higher, though at these pressures up to 5 rings are used on the atmospheric side. Alternatively, if only one ring is used it has to be centered with tilting pads. This has the additional advantage that it increases shaft damping. There is always some barrier liquid entering into the gas space of the machine (litres/hour) which has to be collected, degassed and returned to a reservoir. Further, to prevent any oil creeping along the shaft some gas must flow towards the seal and then be taken to the seal oil traps. The rings themselves are usually made of steel with a white metal lining. Each ring sits in a stationary case.

Leakage around the outside of the ring is prevented by an 'O' ring seal. Floating rings do provide sealing at standstill provided the seal oil flow is maintained. Usually the floating rings are self centering. Some high pressure seals, however, are centred with the help of small auxiliary tilting pad bearings.

Floating ring seals need a constant supply of oil at a pressure higher than the gas which is to be sealed. The equipment needed to supply seal oil, pumps, filters and overhead tanks, as well as seal oil traps and degassing vessels, is bulky and costly.

14.3.3 *Mechanical seals operating on gas*

In recent years mechanical seals working with a gas rather than liquid film have been developed. Fine grooves (microns deep) 'pump' gas into the space between the faces and so generate a film which keeps them apart. Leakage is small, about 0.5 l/s of free gas for a 100 mm seal at a pressure of 50 bar. Maximum operating pressure is above 100 bar. Another advantage of gas seals is their low power consumption, about 1 percent of that of oil flooded seals. It is usual to install two seals in tandem to allow the leakage to be taken to a stack under normal operation as well as to provide a back up in case of the main seal failing.

As in all mechanical seals there is no contact between the faces while operating and therefore no wear. While gas seals are expensive, they eliminate the need for high pressure oil systems. Gas seals provide sealing at standstill.

All three types of positive seals can usually be inspected and changed without opening the casing of a compressor.

14.4 Other seals

14.4.1 *Contact seals*

Another method of keeping clearances to a minimum is the use of split rings which are allowed to touch the rotating shaft (see Fig. 14.5). The most common material for such rings is carbon. Pairs of rings are placed in holders such that their splits are not aligned. The rings are kept in contact with the shaft by

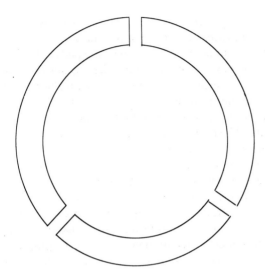

Fig 14.5   Split ring seal

springs placed around their outside. This method of sealing is particularly favoured for small steam turbines, condensate providing the necessary lubricant between shaft and seal. Note that such seals can only be used if the gas handled, the shaft and the seal material are compatible. In particular dry gases are unsuitable for this method of shaft sealing—excessive wear or even pick up is likely to occur.

Chapter 15

Inspection and Testing

15.1 Introduction

Minor machines, e.g. small fans and instrument air compressors, are designed and manufactured to the maker's standard and are frequently bought from stock without any special design or test requirements.

Major machines are usually not available as a standard design but are made to order. Materials, and also allowable stresses for the more critical components, are selected and agreed between the vendor and purchaser at the time of order or during a co-ordination meeting. It is at this stage that the two parties have to agree on the extent of auditing, inspection and testing. On novel types of compressors and machines which require extended tests, the ability to carry out such tests, will influence the choice of vendor.

It is common practice to verify that the vendor adheres to the agreed specification. To achieve this, continuous contact between purchaser and vendor is required during the completion of an order.

A description of the types of tests available is given in the various API specifications (2–6) but in most cases it is up to the purchaser to specify which test is required.

15.2 Design audit

The first stage in this process is for the vendor to submit data, drawings, calculations and material orders for the purchaser's approval. Any deviation to the previously agreed specification should be highlighted by the vendor and should receive special attention. Usually it is possible to accommodate changes, but at times even minor changes can be detrimental. For example, a small change in a material specification can be responsible for major changes in corrosion resistance. Experience of corrosion resistance in particular applications is usually only available to an operator. Such knowledge is rarely available to the vendor.

15.3 Inspection

For major items such as casings, shafts, cylinders, impellers and piston rods, material certificates showing the composition and the requested strength data should be supplied. Pressure containing items such as cylinders and casings should be hydraulically pressure tested to an agreed code. Note that these items are not at present subject to the boiler or vessel code in the UK and therefore need not be retested at specified intervals. However, all vessels such as intercoolers and separators associated with a machine, are subject to the relevant pressure vessel code, e.g. BS 1500 (**27**) or ASME VIII (**28**).

Major items are always inspected by the vendor's personnel. Frequently purchasers will insist that their own inspector or an inspector from one of the major agencies, e.g. Loyds, will also follow the progress of critical items.

15.4 Mechanical test

It is customary for fans and compressors, and specified for some types of machines in API standards, e.g. centrifugal compressors (**3**), to be tested at full speed for a specified period (usually 4 hours) to prove that all parts are free and do not touch or rub. Such running tests are usually followed by a leak test at full pressure. Frequently it is requested that a machine be disassembled after a mechanical test to examine the internals and prove that no contact had occurred. If spare running gear has been supplied it is customary to reassemble with the spare parts and occasionally a second mechanical test is carried out. To save power and to prevent overheating these tests are frequently carried out at a reduced inlet pressure or even with the casing under vacuum. For machines driven by variable speed drivers 'maximum continuous speed' shall be maintained for an hour during the mechanical test.

It is not practical to carry out mechanical tests on larger reciprocating compressors though smaller machines, which do not need a special foundation, are frequently tested.

15.5 Performance test

It is rarely that process machines, other than air compressors, can be tested under full operating conditions in the maker's works. The only gases available in the makers works besides air are frequently Natural gas and sometimes helium mixtures to simulate the density of light gases; further, many machines operate at elevated temperatures.

An exception are refrigeration compressors which are supplied as a complete package together with driver and all associated vessels. Most vendors of these units are equipped to carry out a full test under the normal operating conditions.

The decision to carry out a performance test should be based on the following points:

(a) Is the design novel or is it a 'standard machine' of a type and size the vendor has supplied before? Generally it is easier to increase the size of a given type than to design a machine with a smaller capacity.
(b) How high is the density of the gas? The higher the density the greater the

volume reduction in a stage and the more critical the matching of the various stages. Note however that a gas mixture with a suitable density and κ value may not be available at the test location.

(c) How high is the operating pressure? The higher the pressure the more uncertainty there is in the gas properties and also the greater the chance of roto-dynamic instabilities due to whirl induced in the internal parts of the machine. These effects are not there at lower pressures. If tests are to be carried out at a reduced pressure, this pressure should be specified.

(d) Meaningful performance tests are time consuming and costly. Delivery is usually at a premium.

Note that on large reciprocating machines performance tests are only carried out once they are erected on site.

15.6 String test

String tests require the whole machine to be assembled, together with its proper driver.
There are two distinct parts to such tests:

(a) Do the items fit together? This is a simple test and particularly favoured for multi-casing centrifugal compressors with turbine drivers. When specified it is advantageous to completely erect all minor piping above the base frame.

(b) Do the items operate together? This is a much more difficult test on high power multi-casing machines. Even if the power should be available there rarely is a 'sink' at the maker's works to absorb the heat of compression. A test with the compressor casings operating in a vacuum shows no more than the mechanical test.

Standard codes specifying how to carry out performance tests for fans and compressors are published in API, ISO and other authorities.

15.7 Recommendation

It is difficult to standardise the test requirements. Tests are expensive and, often more important, time consuming. In deciding what test to specify the following should be considered:

(a) Experience of the vendor. Is it a new design or a standard model which is acceptable without special tests?

(b) How close is the final destination from the maker's factory? A machine made in Switzerland and erected in England is 'close' to the manufacturer. One installed in a developing country or on an oil platform is 'far'. The further the place of use the more tests are justified.

(c) How important is the machine, and how does the cost of a day's non-availability compare with the cost of testing?

Chapter 16

Containment Safety

16.1 Introduction

One of the prime requirement of any piece of equipment is that it is safe in use. For fans and compressors that means mainly containment: containment of the gas handled and containment of the fast moving parts of machines in the case of failure. Both these requirements have to be addressed in the design stage. The first step is to recognise that some machines can suffer serious failures. One major user classifies this type of machine as 'critical machines' (**29**). In addition to hazard and operability studies during the design stage, they are subjected to detailed study by the vendors and the user's personnel to evaluate all possible modes of failure as well as how these failures can be avoided or their consequences at least minimised. However, it is not enough to obtain a safe unit in the first place, the unit must be kept in a safe condition during its working life. Regular inspection by experts is needed of parts assessed as likely to deteriorate. To ensure that this is carried out a register of all critical machines on a site should be kept containing details of the steps needed to ensure safety. Contact with the vendor must be maintained throughout the life of the machine so that up to date operating experience can be shared and improvements in operation and protection can be incorporated.

Note that modifications to existing equipment has in the past been the reason behind many failures. If a machine is to be modified the proposed alterations should be investigated to the same extent as for a new machine, operability studies and a fresh hazard analysis are called for.

16.2 Gas containment

16.2.1 *Pressure rating*

Standard design practice requires that all pressure containing parts are so designed that they are safe to operate at the design pressure and that they are tested at a higher pressure before being put into use. To prevent the design pressure from being exceeded relief valves of adequate capacity are fitted. Normally a relief valve on the delivery will also protect the machine though this is not so in the case of twin screw compressors where the internal pressure is dependent not on the pressure in the delivery system but on the pressure in the suction system (see Chapters 7–9).

Attention must be paid to ensure that parts can not be wrongly assembled such as to cause excessive pressures. For example, on reciprocating compressors if an inlet valve (or a delivery valve installed in the reverse direction) is installed in the delivery port, so preventing any flow into the delivery system, excessive pressures will occur inside the cylinder which will usually lead to failure (see Chapter 6). It is therefore essential at the design stage to ensure that such mix-ups can not occur (i.e. they are designed out). Another source of excessive pressure or force are the result of conditions not foreseen, for example, liquid slugs in positive displacement machines designed to handle gas. Liquid is non-compressible and enormous forces can be generated leading to mechanical damage and loss of containment. It is therefore necessary that, if there is any possibility of liquid in the system, adequate separators with level alarms are installed.

Fires and explosions inside the delivery of lubricated air compressors are another source of danger. The heat generated can weaken vessels and pipework so that they can no longer contain the pressure and explosion can rip whole systems apart. Another source of major failures are crankcase explosions. A hotspot inside a crankcase can, if air is suddenly admitted, ignite an oil/air mixture. Such an explosion can develop a pressure as high as 8 bar. If the crankcase can not withstand this pressure, pressure relieve valves are required (31).

16.2.2 Leakage

Items are tested to ensure there is no leakage from pressure containing parts. However, leakage does occur at glands and restrictive seals. Such leakage, if toxic or flammable, should be vented to a safe location. Steam, nitrogen or air may be needed to purge such leaks.

16.2.3 Failure

All designs should be studied for likely sources of failure. For example, any fasteners inside the gas containing part of machines, particularly reciprocating machines, should be safely secured and designed in such a way that even if they should fail debris does not enter into a cylinder. Special consideration must be given to ensure containment in the case of failure. For example, in case of a loss of barrier oil (which itself should initiate a trip) it is conceivable to lose all the white metal in floating ring seals (see Chapter 14), so increasing the very fine clearance normally sealed with oil to a much larger one no longer sealed with the barrier liquid. This would result in a large gas flow into the oil system and this system, coupling guards, drain lines and oil tank, etc., must therefore be fitted with a relieve device such that no excessive pressures occur.

Automatic isolation valves can be used to minimise the escape of gas. Such valves can be actuated by sensors indicating abnormal conditions. Note that such valves do not prevent a gas escape but only minimise it.

Repairs and modifications need to be carefully reviewed to ensure that the design limits are not exceeded.

16.3 Parts containment

High speed machines have parts which, due to the high speed, store considerable amounts of energy. Under normal conditions this kinetic energy is stored within the moving item though when this fails the stored energy becomes available to propel the item or part of it, and cause damage, or in the case of low pressure machines, e.g. fans, to penetrate the pressure casing. There is usually no case for designing casings to withstand such damage if you ensure by testing, e.g. overspeed, and inspection of material and fabrication, that the rotating element is sound. Note that in medium and high pressure machinery with casings designed to withstand more than about 30 bar there is not enough energy in the rotating parts to penetrate the casing in case of failure. Coupling guards, however, are only designed to prevent oil leaks and access to moving parts. They are never capable of containing a burst coupling. In high pressure reciprocating machines, having pistons with tailrods, fractures have occurred in the latter and the detached tailrod has been propelled like a projectile. To ensure its containment a tailrod catcher is required which brings the broken part to rest within the machine.

16.4 Monitoring

16.4.1 *Inspections*

Inspection of machines differs significantly from the inspection of vessels and boilers. On the latter load changes that can lead to fatigue occur at infrequent intervals, so a routine inspection for cracks of such items at annual or biannual intervals is sensible as propagation of cracks is slow. Further regular inspection of boilers and pressure vessels is a legal requirement. On fast running machines however—say 160 r/s (10 000 rpm)—10 000 000 cycles happen in 1000 minutes or just over 15 hours. Even slow running reciprocating compressors subject some parts to 10 million load reversals in about 20 days. Routine inspection of such parts for cracks is not feasible. In the design and manufacture of such parts special measures to keep stresses low and avoid any stress raisers must be taken. Inspection during manufacture is needed to avoid internal flaws. Such inspection is also required after parts are reconditioned.

Frequent inspections for cracks are therefore called for. However, during operation the 'health' (does it leak oil or gas, vibrate or operate too hot?, etc.) and the performance of a machine should be monitored continuously. Any shortcoming of either 'health' or performance should be investigated, the suspect item inspected and rectified.

Conversely, if a machine is healthy and there is no change in performance there is no need for inspection. Taking the machine apart might even cause some damage. Routine inspection should focus on wear.

Note that legislation in some countries requires that machines on certain duties, which have the potential for specific hazards, have a scheme of inspection prepared and records of the inspection retained.

16.4.2 *Monitoring*

To establish the performance of a machine, monitoring is needed. This usually means measuring inlet and delivery pressure and temperature of every stage. With these data the polytropic efficiency can be calculated. Changes in efficiency need further investigation. Flow measurement with sufficient accuracy is rarely available. To monitor 'health' many sensors are available. All machines should regularly be visually inspected to see if there are any leaks. Oil condition and vibration levels should be recorded and compared with those of the newly installed unit. For slow running machines and gears external detectors are the preferred choice of vibration monitoring. For fast running machines proximeters provide more information and their use has resulted in a significant improvement in reliability.

16.4.3 *Trips*

To prevent consequential damage suitable sensors, e.g. oil pressure, axial position and shock or vibration, should be installed to trip a unit in case of trouble. Alarming is of little use as consequential damage usually occurs rapidly, within a fraction of a second, after a primary failure.

Appendices

Appendix I

I.1 Mechanical data sheets

The preliminary task of the machine engineer is to use the information provided on the process data sheet to select a suitable type of machine and to issue a mechanical data sheet. This will form the basis of enquiries for quotes. Note that the mechanical data sheet will initially only contain the principal data, rate and gas properties as well as suction and delivery conditions. Additional information will be provided by vendors when they quote for a specific machine. Once a selection is made the remaining questions raised by the mechanical data sheet should be addressed and settled in a meeting between the purchaser and the vendor (co-ordination meeting).

The most commonly used data sheets are those issued by API and Europneu or data sheets based on the above. A data sheet for centrifugal compressors based on API is given as a sample.

Data sheet 1

	Project No.	Reg'd No.	Equipment No.
		Sheet of Rev	

CENTRIFUGAL COMPRESSOR DATA SHEET

JOB NO. _____ ITEM NO. _____
PURCH. ORDER NO. _____ DATE _____
REQUISITION NO. _____
INQUIRY NO. _____
PAGE __1__ OF __6__ BY _____

APPLICABLE TO: ○ PROPOSAL   ○ PURCHASE   ○ AS BUILT   DATE _____   REVISION _____
FOR _____   UNIT _____
SITE _____   SERIAL NO. _____
SERVICE _____   NO. REQUIRED _____
MANUFACTURER _____ ○ MODEL _____   DRIVER _____
NOTE: INFORMATION TO BE COMPLETED: ○ BY PURCHASER;   ☐ BY MANUFACTURER.

OPERATING CONDITIONS

(ALL DATA ON PER UNIT BASIS)	NORMAL	RATED	OTHER CONDITIONS			
			A	B	C	D
○ GAS HANDLED (ALSO SEE PAGE ___)						
○ Nm³/s						
○ WEIGHT FLOW kg/MIN (WET) (DRY)						
INLET CONDITIONS:						
○ PRESSURE (bara)						
○ TEMPERATURE (C)						
○ RELATIVE HUMIDITY (%)						
○ MOLECULAR WEIGHT (M)						
☐ $C_P C_V (K_1)$ OR (K_{AVG})						
☐ COMPRESSIBILITY (Z_1) OR (Z_{AVG})						
☐ INLET VOLUME (m³/s)						
DISCHARGE CONDITIONS:						
○ PRESSURE (bara)						
☐ TEMPERATURE (C)						
☐ $C_P C_V (K_2)$ OR (K_{AVG})						
☐ COMPRESSIBILITY (Z_2) OR (Z_{AVG})						
☐ REQUIRED (ALL LOSSES INCL)						
☐ SPEED (RPM)						
☐ ESTIMATED SURGE, m³/s (AT SPEED ABOVE)						
☐ POLYTROPIC HEAD (m)						
☐ POLYTROPIC EFFICIENCY (%)						
○ GUARANTEED POINT						
☐ PERFORMANCE CURVE NO.						

PROCESS CONTROL:

METHOD   ○ BYPASS FROM _____ TO _____
     ○ ANTISURGE BYPASS: ○ MANUAL ○ AUTOMATIC
     ○ SUCTION THROTTLING FROM _____ TO _____
     ○ SPEED VARIATION FROM _____ TO _____
     ○ OTHER _____
SIGNAL   ○ SOURCE _____
     ○ TYPE _____
     ○ RANGE: FOR PNEUMATIC CONTROL _____ rps @ _____ bar g: & _____ rps @ _____ bar g _____
     OTHER _____

SERVICE: ○ CONTINUOUS   ○ INTERMITTENT   ○ STAND BY

Data sheet 2

		Project No.	Req'd No.	Equipment No.
			Sheet of Rev	

CENTRIFUGAL COMPRESSOR DATA SHEET

JOB NO. _____ ITEM NO. _____
PAGE ____2____ OF ___6___ BY _____
DATE _____ REVISION _____

OPERATING CONDITIONS, CONT'D

GAS ANALYSIS: ○ MOL % ○ _____	NORMAL	RATED	OTHER CONDITIONS A	B	C	D	REMARKS
	MW						
AIR	28.966						
OXYGEN	32.000						
NITROGEN	28.016						
WATER VAPOR	18.016						
CARBON MONOXIDE	28.010						
CARBON DIOXIDE	44.010						
HYDROGEN SULFIDE	34.076						
HYDROGEN	2.016						
METHANE	16.042						
ETHYLENE	28.052						
ETHANE	30.068						
PROPYLENE	42.078						
PROPANE	44.094						
i - BUTANE	58.120						
n - BUTANE	58.120						
i - PENTANE	72.146						
n - PENTANE	72.146						
HEXANE PLUS							
TOTAL							
AVG. MOL. WT							

LOCATION:
○ INDOOR   ○ HEATED   ○ UNDER ROOF
○ OUTDOOR   ○ UNHEATED   ○ PARTIAL SIDES
○ GRADE   ○ MEZZANINE   ○ _____
○ ELECTRICAL AREA CLASS   GR. _____ DIV. _____
○ WINTERIZATION REQD.   ○ TROPICALIZATION REQD.

SITE DATA:
○ ELEVATION _____ m.   BAROMETER _____ :bar a
○ RANGE OF AMBIENT TEMPS.

	DRY BULB	WET BULB
SITE RATED' c		
NORMAL c		
MAXIMUM c		
MINIMUM c		

UNUSUAL CONDITIONS: ○ DUST   ○ FUMES
○ OTHER _____

NOISE SPECIFICATIONS:
○ APPLICABLE TO MACHINE:
   SEE SPECIFICATION_____
○ APPLICABLE TO NEIGHBORHOOD.
   SEE SPECIFICATION_____
ACOUSTIC HOUSING: ○ YES   ○ NO

APPLICABLE SPECIFICATIONS:
API 617, CENTRIFUGAL COMPR. FOR GEN. REFINERY SERVICES

PAINTING:
○ MANUFACTURER'S STD.
○ OTHERS _____

SHIPMENT:
○ DOMESTIC   ○ EXPORT   ○ EXPORT BOXING REQD.
○ OUTDOOR STORAGE MORE THAN 6 MONTHS

REMARKS: _____

Data sheet 3

	Project No.	Req'd No.	Equipment No.
		Sheet of Rev	

CENTRIFUGAL COMPRESSOR DATA SHEET

JOB NO. _____   ITEM NO. _____
PAGE ___3___ OF ___6___ BY _____
DATE _____   REVISION _____

CONSTRUCTION FEATURES

☐ **SPEEDS:**

MAX. CONT. _____ rps   TRIP _____ rps

MAX. TIP SPEEDS: _____ FPS @ RATED SPEED

_____ FPS @ MAX. CONT. SPEED

☐ **LATERAL CRITICAL SPEEDS:**

FIRST CRITICAL _____ rps

DAMPED _____ UNDAMPED _____

MODE SHAPE _____

SECOND CRITICAL _____ rps

DAMPED _____ UNDAMPED _____

MODE SHAPE _____

THIRD CRITICAL _____ rps

DAMPED _____ UNDAMPED _____

MODE SHAPE _____

FOURTH CRITICAL _____ rps

DAMPED _____ UNDAMPED _____

MODE SHAPE _____

LATERAL CRITICAL SPEED — BASIS:

☐ DAMPED UNBALANCE RESPONSE ANALYSIS

☐ SHOP TEST

☐ OTHER TYPE ANALYSIS

☐ **TORSIONAL CRITICAL SPEEDS:**

FIRST CRITICAL _____ rps

SECOND CRITICAL _____ rps

THIRD CRITICAL _____ rps

FOURTH CRITICAL _____ rps

☐ **VIBRATION:**

ALLOWABLE TEST LEVEL _____ μm
(PEAK TO PEAK)

☐ **ROTATION, VIEWED FROM DRIVEN END:**

☐ **CASING:**

MODEL _____

CASING SPLIT _____

MATERIAL _____

THICKNESS (mm) _____ CORR. ALLOW (mm) _____

MAX. WORK. PRESS. _____ bar g   MAX DESIGN PRESS. _____ bar g

TEST PRESS (bar g): HELIUM _____ HYDRO _____

MAX. OPER. TEMP. _____ C   MIN. OPER. TEMP. _____ C

MAX. NO. OF IMPELLERS FOR CASING _____

MAX. CASING CAPACITY (m³/s) _____

RADIOGRAPH QUALITY   ◯ YES _____   ◯ NO _____

CASING SPLIT SEALING _____

☐ **DIAPHRAGMS:**

MATERIAL _____

☐ **IMPELLERS:**

NO. _____ DIAMETERS _____

NO. VANES EA IMPELLER _____

TYPE (OPEN, ENCLOSED, ETC.) _____

TYPE FABRICATION _____

MATERIAL _____

MAX. YIELD STRENGTH (bar) _____

BRINNEL HARDNESS: MAX. _____   MIN. _____

SMALLEST TIP INTERNAL WIDTH (mm) _____

MAX. MACH NO. @ IMPELLER EYE _____

MAX. IMPELLER HEAD @ RATED SPEED (M) _____

☐ **SHAFT:**

MATERIAL _____

DIA. @ IMPELLERS (mm) _____   DIA. @ COUPLING (mm) _____

SHAFT END: ☐ TAPERED   ☐ CYLINDRICAL

MAX. YIELD STRENGTH (bar) _____

☐ **BALANCE PISTON:**

MATERIAL _____ AREA _____ (mm²)

FIXATION METHOD _____

SHAFT SLEEVES:

◯ AT INTERSTG. CLOSE CLEAR. PTS. ☐ MATL _____

◯ AT SHAFT SEALS _____ ☐ MATL _____

_____

☐ **LABYRINTHS:**

INTERSTAGE

TYPE _____ MATERIAL _____

BALANCE PISTON

TYPE _____ MATERIAL _____

SHAFT SEALS:

☐ TYPE _____

◯ SEAL SYSTEM TYPE _____

◯ SETTLING OUT PRESSURE _____

◯ INNER OIL LEAKAGE GUAR. (GAL/DAY/SEAL) _____

◯ TYPE BUFFER GAS _____

☐ BUFFER GAS FLOW (PER SEAL): _____

NORMAL: _____ kg/s @ _____ bar Δ P _____

MAX.: _____ kg/s @ _____ bar Δ P _____

◯ BUFFER GAS REQUIRED FOR:

☐ START-UP

☐ AIR RUN-IN

☐ OTHER _____

☐ BUFFER GAS CONTROL

SYSTEM SUPPLIED BY _____

☐ **BEARING HOUSING CONSTRUCTION:**

TYPE (SEPARATE, INTEGRAL) _____ SPLIT _____

MATERIAL _____

Data sheet 4

	Project No.	Req'd No.	Equipment No.
		Sheet of Rev	

	JOB NO. _____ ITEM NO. _____
CENTRIFUGAL COMPRESSOR DATA SHEET	PAGE __4__ OF __6__ BY _____
	DATE _____ REVISION _____

CONSTRUCTION FEATURES, CONT'D

☐ **RADIAL BEARINGS:**

TYPE _____ SPAN (mm) _____

AREA (mm²) _____ LOADING (bar) ACT. _____ ALLOW. _____

CENTER PIVOT _____

OFFSET PIVOT _____

% _____

PAD MATERIAL _____

TYPE BABBITT _____

BABBITT THICKNESS _____

☐ **THRUST BEARING:**

LOCATION _____ TYPE _____

MFR _____ AREA (mm²) _____

LOADING (bar) ACTUAL _____ ALLOWABLE _____

GAS LOADING (kg) _____ CPLG. SLIP LOAD (kg) _____

CPLG. COEFF. FRICT. _____ CPLG. GEAR PITCH DIA. (mm) _____

BAL. PISTON COMPENSATING LOAD _____

CENTER PIVOT _____

OFFSET PIVOT _____

% _____

PAD MATERIAL _____

TYPE BABBITT _____

BABBITT THICKNESS _____

☐ **MAIN CONNECTIONS:**

	SIZE	ANSI RATING	FACING	POSITION	FLANGE VEL. FPS
INLET					
DISCHARGE					

☐ **ALLOWABLE PIPING FORCES AND MOMENTS:**

	INLET		DISCHARGE			
	FORCE N	MOMT Nm	FORCE N	MOMT Nm	FORCE N	MOMT Nm
AXIAL						
VERTICAL						
HORIZ 90						

	FORCE N	MOMT Nm	FORCE N	MOMT Nm	FORCE N	MOMT Nm
AXIAL						
VERTICAL						
HORIZ 90						

☐ **OTHER CONNECTIONS:**

SERVICE:	NO	SIZE	TYPE
LUBE-OIL INLET			
LUBE-OIL OUTLET			
SEAL-OIL INLET			
SEAL-OIL OUTLET			
CASING DRAINS			
STAGE DRAINS			
VENTS			
COOLING WATER			
PRESSURE			
TEMPERATURE			
PURGE FOR:			
BRG. HOUSING			
BETWEEN BRG. & SEAL			
BETWEEN SEAL & GAS			
SOLVENT INJECTION			

VIBRATION DETECTORS:

○ TYPE _____ ☐ MODEL _____

○ MFR _____

○ NO. AT EACH SHAFT BEARING _____ TOTAL NO. _____

○ OSCILLATOR-DETECTORS SUPPLIED BY _____

○ MFR _____ ☐ MODEL _____

○ MONITOR SUPPLIED BY _____

○ LOCATION _____ ENCLOSURE _____

○ MFR _____ ☐ MODEL _____

☐ SCALE RANGE _____ ○ ALARM ☐ SET @ _____ μm

○ SHUTDOWN: ☐ SET @ _____ μm ○ TIME DELAY _____ SEC.

AXIAL POSITION DETECTOR:

○ TYPE _____ ☐ MODEL _____

○ MFR _____ ○ NO. REQUIRED _____

○ OSCILLATOR-DEMODULATOR SUPPLIED BY _____

○ MFR _____ ☐ MODEL _____

○ MONITOR SUPPLIED BY _____

○ LOCATION _____ ENCLOSURE _____

○ MFR _____ ☐ MODEL _____

☐ SCALE RANGE _____ ○ ALARM ☐ SET @ _____ μm

○ SHUTDOWN: ☐ SET @ _____ μm ○ TIME DELAY _____ SEC

COUPLINGS:

	DRIVER-COMP. OR DRIVER-GEAR	GEAR-COMP
○ MAKE		
☐ MODEL		
○ LUBRICATION		
○ MOUNT CPLG. HALVES		

Data sheet 5

Project No.	Reg'd No.	Equipment No.
	Sheet of Rev	

CENTRIFUGAL COMPRESSOR DATA SHEET

JOB NO. _____ ITEM NO. _____
PAGE __5__ OF __6__ BY _____
DATE _____ REVISION _____

CONSTRUCTION FEATURES, CONT'D

COUPLINGS, CONT'D

	DRIVER-COMP OR DRIVER-GEAR	GEAR-COMP
○ SPACER REQD.		
○ LIMITED END FLOAT REQD.		
○ IDLING ADAPTOR REQD.		
□ CPLG. RATING (HP/100 RPM)		
□ KEYED (1) OR (2); OR HYDR. FIT		

BASEPLATE & SOLEPLATES:

SOLEPLATES FOR ○ COMPRESSOR ○ GEAR ○ DRIVER

BASEPLATE:

○ COMMON (UNDER COMP., GEAR & DRIVER)
○ UNDER COMP. ONLY ○ OTHER _____
○ DECKED WITH NON-SKID DECK PLATE ○ OPEN CONSTR.
○ DRIP RIM ○ WITH OPEN DRAIN
○ HORIZ. ADJUSTING SCREWS FOR EQUIPMENT
○ SUITABLE FOR POINT SUPPORT
○ SUITABLE FOR PERIMETER SUPPORT
○ STAINLESS SHIMS: THICKNESS _____
□ GROUTING: TYPE _____

SHOP INSPECTION AND TESTS:

	REQD.	WITNESS	OBSERVED
SHOP INSPECTION	○	○	○
HYDROSTATIC	○	○	○
HELIUM LEAK	○	○	○
MECHANICAL RUN	○	○	○
MECH. RUN SPARE ROTOR	○	○	○
FIT IN SPARE ROTOR	○	○	○
PERFORMANCE TEST (GAS) (AIR)	○	○	○
COMP. WITH DRIVER	○	○	○
COMP. LESS DRIVER	○	○	○
USE SHOP LUBE & SEAL SYS.	○	○	○
USE JOB LUBE & SEAL SYS.	○	○	○
USE SHOP VIBRATION PROBES. ETC.	○	○	○
USE JOB VIB. & AXIAL DISP. PROBES. OSCILLATOR-DETECTORS & MONITOR	○	○	○
PRESSURE COMP. TO FULL OPER. PRESS	○	○	○

DISASSEMBLE-REASSEMBLE COMP.
AFTER TEST   ○   ○   ○
CHECK BRGS & SEALS AFTER TEST   ○   ○   ○
NOISE LEVEL TEST   ○   ○   ○
RESIDUAL ELECTRICAL/MECH. RUNOUT   ○   ○   ○
_____ ○   ○   ○

□ **WEIGHTS (kg):**
COMPR. _____ GEAR _____ DRIVER _____ BASE _____
ROTORS: COMPR. _____ DRIVER _____ GEAR _____
COMPR. UPPER CASE _____
L.O. CONSOLE _____ S.O. CONSOLE _____
MAX. FOR MAINTENANCE (IDENTIFY) _____
TOTAL SHIPPING WEIGHT _____

□ **SPACE REQUIREMENTS (METRES)**
COMPLETE UNIT: L _____ W _____ H _____
L.O. CONSOLE: L _____ W _____ H _____
S.O. CONSOLE: L _____ W _____ H _____

MISCELLANEOUS:

□ RECOMMENDED STRAIGHT RUN OF PIPE DIAMETERS BEFORE SUCTION _____
○ VENDOR'S REVIEW & COMMENTS ON PURCHASER'S PIPING & FOUNDATION
○ OPTICAL ALIGNMENT FLATS REQUIRED ON COMPRESSOR, GEAR & DRIVER
○ PROVISION FOR WATER WASHING BEFORE OPENING CASING BY _____
○ TORSIONAL ANALYSIS REPORT REQUIRED
_____
_____
_____
_____
_____
_____
_____

REMARKS: _____
_____

Data sheet 6

CENTRIFUGAL COMPRESSOR DATA SHEET

Project No.	Reg'd No	Equipment No.
	Sheet of Rev	

JOB NO.
PAGE 6 OF 6
DATE

ITEM NO.
BY
REVISION

UTILITIES

UTILITY CONDITIONS:

STEAM

	DRIVERS		HEATING	
INLET MIN	bar g	C	bar g	C
NORM	bar g	C	bar g	C
MAX	bar g	C	bar g	C
EXHAUST. MIN	bar g	C	bar g	C
NORM	bar g	C	bar g	. C
MAX	bar g	C	bar g	C

ELECTRICITY:

	DRIVERS	HEATING	CONTROL	SHUTDOWN
VOLTAGE				
HERTZ				
PHASE				

COOLING WATER:

TEMP. INLET	C	MAX RETURN C
PRESS NORM	bar g	DESIGN bar g
MIN RETURN	bar g	MAX ALLOW Δ P bar
WATER SOURCE		

INSTRUMENT AIR:

MAX PRESS	bar g	MIN PRESS bar g

□ **TOTAL UTILITY CONSUMPTION:**

COOLING WATER	l/s
STEAM, NORMAL	kg
STEAM, MAX.	kg
INSTRUMENT AIR	cm³/hr
HP (DRIVER)	kw
HP (AUXILIARIES):	kw

REMARKS:

Appendix II

II.1 Properties of gas mixtures

It is usual to give gas composition in % volume at 0 C and 1 bar. The basis for selecting machines design is, however, based on other parameters namely, gas density, compressibility and ratio of specific heats of the gas or gas mixture handled. In this appendix the conversion from one to the other is outlined in two sections:

(1) Gas properties
   Density and ratio of specific heats and deviation from ideal gas behaviour, particularly at high pressures.
(2) Humidity.

The density of a gas 'ρ_0' at normal conditions can be calculated from its molar mass (molecular weight) 'M_w'. 1 kg mole of gas has a volume of 22.4 m^3, and its density 'ρ_0' is $M_w/22.4$ kg/m^3. The density ρ at operating conditions is calculated, using the ideal gas law, i.e. $pv = RT$, to be

$$\rho = \rho_0 p T_0 / p_0 T \tag{II.1}$$

At elevated pressures the behaviour of gases deviates from this law. A simple, and adequate way for a preliminary selection, to allow for this deviation is the use of the compressibility factor 'z'. Compressibility factors for individual gases can be found in reference (**18**) and some data books. To find the compressibility, particularly of a mixture, the use of universal graphs using reduced pressure and temperature as parameters is described below. The reduced pressure and temperature are the ratios between actual and critical pressure and temperature of the gas/gas mixture. Figs II.1 and 2 give the compressibility depending on reduced pressure and temperature (**31**).

For a mixture the molar mass M_w is

$$M_w = \Sigma r_i M_i \tag{II.2}$$

where M_w the molar mass of the mixture

r_i the volumetric percentage of component i

M_i the molecular mass of component i.

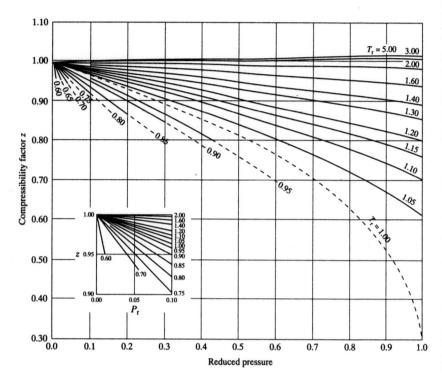

Fig AII.1   Generalized compressibility coefficient for gases depending on reduced pressure and temperature. (Source: Nelson, L.C., Obert, E.F., *Trans 76*, 1954, 1057)

The critical pressure p_c of a mixture is

$$p_c = \Sigma \, r_i p_{ci} \tag{II.3}$$

The critical temperature

$$T_c = \Sigma \, r_i T_{ci} \tag{II.4}$$

The reduced pressure p_r, and reduced temperature T_r are

$$p_r = p/p_c \tag{II.5}$$

$$T_r = T/T_c \tag{II.6}$$

Knowing the compressibility factor the ideal gas equation becomes:

$$pv = zRT \tag{II.7}$$

the corrected density therefore becomes

$$\rho = \rho_0 p T_0 / z p_0 T \tag{II.8}$$

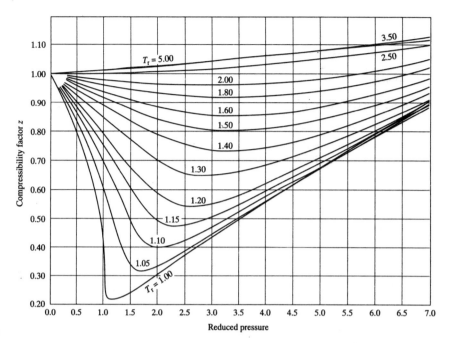

Fig AII.2 Generalized compressibility coefficient for gases depending on reduced pressure and temperature. (Source: Nelson, L.C., Obert, E.F. *Trans ASME*, 1954, 76, 1057)

Note it is more correct to include z_0 too, i.e.

$$\rho = \rho_0 z_0 p T_0 / z p_0 T \tag{II.9}$$

but z_0 is nearly always unity and is therefore neglected.

The ratio of specific heats κ of the mixture can be obtained from the specific heat at constant pressures for each component as follows

$$M_w c_p = \Sigma\, r_i M_{wi} c_{pi} \tag{II.10}$$

$$\kappa = M_w c_p / (M_w c_p - R) \tag{II.11}$$

The speed of sound u_a in a gas can be calculated from the molar mass M_w and the ratio of specific heats κ

$$u_a = \mathbf{V}(100 \kappa z T R / M_w) \tag{II.12}$$

Methods for calculating gases at high pressures
Under some conditions, e.g. high pressures or pressure ratios, the mixing formulae given above do not hold: some gases in a mixture do not behave in accordance with their partial pressure but are influenced by the total pressure.

Various ways of calculating the properties under these conditions have been developed. Their inclusion is not required in this guide though two useful references are (32) and (33).

II.2 Water

The presence of water in a gas has two significant effects on machines:

(a) A process usually requires 'dry gas rate', so the fan/compressor needs to have a suction rate to cope not only with the air/gas but also with the water vapour. Further the presence of water vapour will alter the density of the mixture.

(b) On multi-stage machines water, and other condensables, will be separated after intercoolers/aftercoolers once saturation is reached. The amount separated needs to be known to size catchpots, etc.

If the water content is given as percentage volume or molar mass the density is calculated for a gas mixture with water as one of the components. If the water content is given as percentage a method is given below. When given as a wet bulb temperature it is best to first calculate the % water vapour. Tables correlating wet bulb and partial pressure can be found in (34) and other handbooks. To calculate the amount of water separated after compression and cooling it is simplest to use percentage saturation. Table II.1 gives saturation pressure against temperature.

Table II.1

Temperature (C)	Saturation pressure (mbar)	Temperature (C)	Saturation pressure (mbar)
0	6.12	35	56.1
5	8.71	40	73.7
10	12.2	45	95.6
15	17.0	50	123.0
20	23.3	60	199.0
25	31.6	70	312.0
30	42.3	80	474.0

The dry volume Q_d can be obtained as follows:

$$Q_d = Q(p - p_{water})/p \qquad (II.13)$$

where p_{water} is the actual water vapour pressure.

The dry mass can be calculated by multiplying this volume with the dry density. Note that p_{water} is the product of saturation pressure p_s and percentage saturation φ.

To calculate if saturation is reached after compression and cooling proceed as follows.

Let p_1, T_1, p_{s1}, φ_1 and p_2, T_2, p_{s2}, respectively, be the conditions at inlet and after the stage cooler (conditions at the inlet to the next stage), where p_{si} is the saturation pressure at the temperature T_1 and φ_1 the percentage saturation at condition 1.

The mass of water vapour per unit volume at the inlet is

$$m_{water1} = 273 \times 0.8 p_{s1}\varphi_1/T_1 p_1 \tag{II.14}$$

The total quantity of water is $Q_1 m_{water1}$

The maximum mass of water per unit volume that can be carried under condition 2, i.e. when saturation is reached is:

$$m_{water2} = 273 \times 0.8 p_{s2}/T_2 p_2 \tag{II.15}$$

and the total quantity is $Q_2 m_{water2}$

If this is less than the water present at condition 1 some water will condense and should be separated. The quantity to be separated is

$$M_{sep} = Q_1 m_{water1} - Q_2 m_{water2} \tag{II.16}$$

or

$$= Q_1(273 \times 0.8/T_1)(p_{s1}\varphi_1/p_1 - p_1 p_{s2}/p_2^2) \tag{II.16a}$$

Bibliography

(1) Davidson, J. (editor) (1986) *Process pump selection – a system approach.* Mechanical Engineering Publications, London.

(2) American Petroleum Institute Standard API 673: Fans for general refinery services.

(3) American Petroleum Institute Standard API 617: Centrifugal compressors for general refinery services (1988).

(4) American Petroleum Institute Standard API 618: Reciprocating compressors for general refinery services (1986).

(5) American Petroleum Institute Standard API 619: Rotary type positive displacement compressors for general refinery services (1985).

(6) American Petroleum Institute Standard API 672: Packaged, integrally geared centrifugal plant and instrument air compressors for general refinery services.

(7) International Standards Organization, ISO 13 707: Reciprocating compressors for the petroleum and natural gas industries.

(8) International Standards Organization, ISO 10439: Centrifugal compressors for the petroleum and natural gas industries.

(9) International Standards Organization, ISO 10440: Rotary compressors for the petroleum and natural gas industries.

(10) Davidson, J. (editor) (1994) *The reliability of mechanical systems* (Second Edition), Mechanical Engineering Publications, London.

(11) Eck, B. (1973) *Fans – Design and operation of centrifugal, axial and crossflow fan*, Pergamon Press, New York.

(12) British Standard for fan testing BS 848 para 9.

(13) *Acceptance and performance testing of fans*, VDI 2044.

(14) Code 91, Standard for the installation of blower and exhaust systems for dust, stock and vapor removal or conveying, National Fire Protection Association, USA.

(15) *Material requirements, sulfide stress cracking resistant metallic materials for oil field equipment*, National Association of Corrosion Engineers (NACE) MR-01-75.

(16) NEMA SM23, The National Electrical Manufacturers Association.

(17) Stepanoff, A. J. (1955) *Turboblowers: Theory, design, and application of centrifugal and axial flow compressors and fans*, John Wiley/Chapman & Hall, New York.

(18) Froelich, F. (1961) *Kolben Verdichter: Thermodynamische Grundlagen Berechnungen Konstruction und Betriebsverhalten*, Springer Verlag, Berlin.

(19) American Petroleum Institute Standard API 614: Lubrication, shaft-sealing, and control-oil systems for special purpose applications (1992).

(20) Arbon, I. M. (1994) *The design and application of rotary twin shaft compressors in the oil and gas process industry*, Mechanical Engineering Publications, London.

(21) *VGB Richtlinien für Kesselspeisewasser, Kesselwasser und Dampf von Wasserohrkresseln der Druckstufen ab 64 bar*, (1980), VGB Kraftwerk-technik, pp. 793–800.

(22) AGMA American Gear Manufacturers Association.

(23) Neal, M. J., Needham, P. and Horrell, R. (1991) *Couplings and shaft alignment*, Mechanical Engineering Publications, London.

(24) Byrne, R. C. Senior (1988) Tubular Exchanger Manufacturing Association (TEMA) (Seventh Edition).

(25) BSI 4231 (ISO 3448): Classification for viscosity grades of industrial liquid lubricants, British Standard Institution.

(26) Summers-Smith, D. J. (1988) *Mechanical seal practice for improved performance*, Mechanical Engineering Publications, London.

(27) BSI 5500: Fusion welded unfired pressure vessels.

(28) Boiler and Presssure Vessel Code Section VIII. ASME, USA.

(29) Rayner, K. G. and Carick, H. B. (1993) Machinery safety – registration of critical machines. Fluid machinery for the oil, petrochemical and related industries, *Proc. Instn. Mech. Engrs*, 263–268.

(30) Wilson, K. G. *et al.* (1969–70) Compressors: what the user wants and why. Industrial reciprocating and rotary compressors design and operational problems, *IMechE Proceedings*, **184**, Part 3R.

(31) Nelson, L. C. and Oberte, E. F. (1954) *Generalised P-V-T Properties of gases. Trans ASME*, **76**, 1057.

(32) Starling, K. E. (1973) Fluid thermodynamic properties of light petroleum systems, Gulf Publishing Company, USA.

(33) Soave, G. (1922) *Chemical Engng Sci.*, **27**.

(34) *Chemical Engineers Handbook* (1941) McGraw Hill, UK.

Index